Diseases of the Endocannabinoid System

Jean-Paul Henri Dedam • Matthew Fogel
Elizabeth Fogel

Diseases of the Endocannabinoid System

Springer

Jean-Paul Henri Dedam, MD
Albuquerque, NM, USA

Elizabeth Fogel, FNP
CannaCore
Albuquerque, NM, USA

Matthew Fogel, FNP
CannaCore
Albuquerque, NM, USA

ISBN 978-3-031-87483-3 ISBN 978-3-031-87484-0 (eBook)
https://doi.org/10.1007/978-3-031-87484-0

This Springer imprint is published by the registered company Springer Nature Switzerland AG
The registered company address is: Gewerbestrasse 11, 6330 Cham, Switzerland

Preface

The Endocannabinoid system is one of the last *known* unexplored frontiers of modern medicine, and unfortunately, many physicians are still unaware of its existence. Research and education on the subject have undoubtably been hindered by our misguided and racist war on drugs. As the era of prohibition slowly fades, and more and more patients discover the benefits of cannabis as a medicine, talk of the Endocannabinoid system has slowly started to slip into our field of view. I have dedicated the last several years to studying this new system and have even gone so far as starting a symptom management clinic at a local cancer center, helping patients to get the most out of this medicine. I have read a lot of books and a lot of articles on the subject, trying to wrap my own mind around it, and have come to only one definitive understanding: it is really, really complicated.

In medicine we like things simple, we are taught Occum's razor very early on. This is the principle that the simplest solution to a problem is probably the right one. We love algorithms and linear, direct causal relationships. We see the world through Newton's eyes. But the Endocannabinoid system isn't Newtonian, it's quantum.

It is incredibly dynamic, changing more quickly than our slow-paced experiments can keep up with. Take, for example, anandamide, one of the primary ligands in the Endocannabinoid system. Unlike other neurotransmitters, it is not stored in vesicles patiently awaiting the call to action. Instead, it is made on demand. Quickly created, it binds to both cannabinoid receptors and exerts its actions, and then just as quickly it is broken down and gone from the system. Its lifespan is measured in milliseconds, manifesting itself almost like a ghost in our traditional experimental models. THC can bind these same receptors and mimic anandamide. Cannabis medicine should be as simple as that, whatever anandamide can do, THC can do it for longer. That would be great, but the reality of the Endocannabinoid system is so much, much more complex.

For starters, the Endocannabinoid system has a lot more players than just anandamide and the two cannabinoid receptors. Anandamide can bind to at least six other non-cannabinoid receptors in the body. These receptors in turn can also be bound by more than just the commonly cited anandamide and 2 AG, the other best-studied Endocannabinoid. Because anandamide is made and broken down so quickly, changes in the activity of the enzymes that create it and the enzymes that break it down can have a dramatic effect on its function as well. The presence and number of these enzymes themselves changes rapidly, as do all of the receptors that anandamide binds. The effect of this binding can also vary widely depending on the influence of other intracellular signaling pathways, existing external conditions, and intercellular interactions. This complicated dance of myriad different constantly changing partners culminating in a specific outcome warranted the coining of a new term: the Entourage effect. This a very not specific term, and is used in a variety of different contexts. It can refer to the effects of various Endocannabinoids and other related lipid messenger molecules like PEA and OEA. It is often also used to refer to how THC interacts with other phytocannabinoids as well as the terpenes also present in the cannabis. In either care, it does serve to underpin the complexity of the Endocannabinoid system and that its function can rarely be reduced to one ligand acting on one receptor at one point in time.

The Endocannabinoid system is never static. It seems designed to constantly balance itself. Its most well-known function is to serve as a feedback inhibitor in the brain, balancing the effect of other neurotransmitters like GABA and glutamate, but also itself. A deficiency in Endocannabinoid function could potently result in disease, allowing other vital pathways to fall out of balance, and undoubtedly plays an under-appreciated role in many supposedly well-understood disease pathways and many other not so well understood ones as well. It has also been shown to have feed-forward effects, sometimes pushing the body out of balance under certain conditions resulting in disease. The purpose of this book is to pull together existing research on the Endocannabinoid system and to shine a light for the first time on a systematic fashion's role in creating, perpetuating, and at times attenuating different disease states.

As you can see, there is no easy linear path to follow here. The Endocannabinoid system functions more like a complicated web of interconnected parts. You cannot change any one part without having a sometimes unexpected change in the system as a whole. This is incredibly challenging for a medical provider used to prescribing medicines that have a single target and a single function. In many ways, cannabis medicine reminds me more of Traditional Chinese medicine, where patterns of dysfunction are more important than a single clear diagnosis of pathology. Patients come to my clinic bearing the heavy burden of chemotherapy side effects. Everything in their systems is out of balance. They can't sleep. They are nervous. They are stressed. They have no energy and no appetite. They have diarrhea and constipation! They have pain, but many either don't find relief from opiate pain medications or simply don't like how those types of pain medications make them feel. They can't stomach taking another pill. Cannabis is not a panacea, but in a single treatment so many patients find relief from so many of their problems. That in and of itself is a testament to how far-reaching and important the Endocannabinoid system is. Across 14 different body systems, we will detail the role as it is currently understood, with the unfortunate but nonetheless exciting reality that much still waits to be discovered. We will see how the ECS is capable of addressing all the symptoms listed above. Then, we will start to investigate the plain fact that such an important system, which does so much heavy lifting across so many body systems, will fail to do what it is supposed to do, and that failure will manifest as disease. We will review existing evidence for the role various Endocannabinoid deficiency and excess states contribute to known diseases. Unfortunately, until better clinical tools exist for accurately measuring ECS tone in living, breathing patients, we will likely need to rely on proxy measures for diagnosis. Once these tools exist though, I have little doubt that a new chapter in medicine will open: Diseases of the Endocannabinoid System!

Albuquerque, NM, USA Jean-Paul Henri Dedam
Albuquerque, NM, USA Matthew Fogel
Albuquerque, NM, USA Elizabeth Fogel

Contents

Endocannabinoid System Basics

1

To grossly overgeneralize, the central role of the endocannabinoid system in the body is to maintain homeostasis. In a likely impossible attempt to keep things "simple," we will start with a description of the classic endocannabinoid retrograde feedback loop on pre- and postsynaptic neurons in the brain. Both cannabinoid receptors are G protein-coupled receptors. In this example, cannabinoid receptor 1 (CB1) is present in the cell membrane of the presynaptic cell. When the presynaptic cell depolarizes, it causes an influx of intracellular calcium ions in the postsynaptic cell. These ions prompt the production of 2-arachidonoylglycerol or 2-AG, an endocannabinoid. 2-AG is synthesized from membrane phospholipids by the enzymes phospholipase C and then diacylglycerol lipase (DAG-L) in the postsynaptic cell. Activation of DAG-L can also be initiated by the presynaptic cell release of glutamate or GABA, the former an excitatory neurotransmitter and the latter an inhibitory neurotransmitter, which then diffuses across the synaptic cleft and activates the postsynaptic cell by binding the glutamate or GABA receptor, respectively [1]. 2-AG is then transported across the cell membrane by a yet unknown transporter protein but also by simple diffusion [2]. 2-AG is a hydrophobic molecule that is not able to easily diffuse in either intracellular or extracellular spaces such as the synapse. A number of fatty acid binding proteins (FABP), such as FABP5 in the case of 2-AG, have been identified that facilitate movement both inside and outside of the cell and are important for efficient signaling [3]. 2-AG then binds and activates CB1 receptor on the presynaptic cell inhibiting calcium channels resulting in short-term suppression of transmission of that neuron. The CB1 receptor is also capable of inhibiting adenylyl cyclase (AC) and decreasing the production of cAMP, a major intracellular signaling molecule. This effect results in long-term suppression of transmission by the presynaptic neuron. This feedback inhibition loop is the mechanism behind cannabinoid treatment of seizures and migraines, both characterized by excess firing of excretory neurons and glutamate release as well as some of their neuroprotective effects by preventing excitotoxicity. The opposite outcome would occur with feedback inhibition of GABA-releasing neurons, as the cells that were

J.-P. H. Dedam et al., *Diseases of the Endocannabinoid System*,
https://doi.org/10.1007/978-3-031-87484-0_1

being inhibited by the presynaptic release of GABA would become free to depolarize. 2-AG is then immediately broken down predominately by the enzyme MAGL within the presynaptic cell, ending the loop. The CB1 receptor is also known to activate the mitogen-activated protein kinase (MAPK) pathway and the PI3K/Akt pathway, both of which are involved in cell survival and death [4].

The body produces two principal endocannabinoids, 2-AG and anandamide (AEA). 2-AG is a full agonist of CB1, while AEA is only a partial agonist. 2-AG is also present in much higher amounts in brain tissues, although both have been found virtually in every tissue of the body. Both also activate CB2 receptor, which is also a membrane-bound G protein. AEA is also produced on demand in a manner similar to 2-AG except by the enzyme NAPE-PLD and quickly broken down by FAAH. COX-2, the cytochrome p450 enzymes, are also capable of processing both endocannabinoids. Because they are created on demand and broken down very quickly, activation/deactivation and gene upregulation/downregulation of either production enzymes, NAPE-PLD and DAG-L, or degradation enzymes, MAGL and FAAH, can cause significant alterations in endocannabinoid levels and therefore represent additional targets for therapeutic intervention. While the membrane transport protein for 2-AG has not been identified, FLAT, a membrane protein that resembles FAAH but does not have an enzyme activity, has been demonstrated to facilitate movement of AEA across cell membranes [5]. AEA can be produced and degraded by blood cells, so significant care needs to be paid to the time between blood draws and measurement of AEA in experimental models. Both endocannabinoids seem to demonstrate circadian variations as well. 2-AG reaches a nadir at 4 am and then steadily climbs to peak around 1 pm before falling again. AEA is less well studied, but appears to peak in the morning and again late at night. The connection between increased AEA levels and exercise has been well studied and is thought be at least one component of the "runner's higher," making anandamide an apt name given its derivation from the Sanskrit word for bliss [6].

While not considered true endocannabinoids because of their lack of binding affinity to the cannabinoid receptors, palmitoylethanolamide (PEA) and oleoylethanolamide (OEA) share a similar synthesis pathway to AEA and interact with other receptors implicated in the endocannabinoid system including PPAR-gamma, TPPV1, and several orphan G proteins, as discussed below. They are therefore known to be involved in metabolism, inflammation, and pain and along with the previously discussed 2-AG and AEA that play an entourage-like role within the ECS, fine-tuning the response of the ECS. For this reason, they represent potential therapeutic targets themselves [7]. Hemopressin, a fragment of the hemoglobin alpha-1 chain and consisting of around 100 amino acids, is the first peptide that has been known to interact with cannabinoid receptors, acting as an inverse agonist of CB1 receptor and a positive allosteric modulator of CB2 [8]. This non-psychoactive peptide cannabinoid or pepcan has been found to have effects on antinociception, appetite, memory, and GI motility in mice. The distribution of this pepcan in mice appears to be mainly in the locus ceruleus, but also the hippocampus and cerebral cortex, with the majority of the pepcan found in circulation coming from chromatin cells of the adrenal cortex [9]. Its release from the adrenals appears to be in response to tissue damage and, given its effect on allosterically boosting CB2 function, likely also serves to boost CB2's tissue protection role [10].

In the brain, CB2 receptor is not present on neurons themselves but microglia and immune cells. In the body at large, it is more often associated with immune cells, exerting important anti-inflammatory functions, although that is certainly not its only role. Its various locations and roles along with CB1 receptor will be discussed at length in every chapter. They are not, however, the only receptors that 2-AG and AEA are capable of binding and therefore not the only receptors of the ECS. We will briefly introduce the other receptors but significantly more details will follow in the chapters on each organ system. Transient receptor potential vanilloid 1 (TPRV1) is best known for being the receptor activated by capsaicin to produce the burning sensation associated with spicy chili. It can also be activated by heat, acid, and AEA but not 2-AG. It is a membrane-bound cation channel that is similar in structure to potassium channels, but many of its functions depend on the movement of calcium ions. It can be sensitized by inflammation but also desensitized by its ligands to illicit lesser or no pain sensations depending on local conditions [11]. Peroxisome proliferator-activated receptor (PPAR) alpha and gamma are transcription factor receptors that when activated by ligands including AEA, 2-AG, PEA, and OEA regulate gene expression within the nucleus of the cell. They are known to be involved in the control of metabolism including insulin sensitivity and are the drug target of the diabetes drug pioglitazone. They can also have an anti-inflammatory role [12]. An increasing number of orphan membrane-bound G protein-coupled receptors are also activated by the ECS, including GPR18, GPR55, and GPR119. GPR18 is a recently described receptor known to be present in the brain stem and lowers BP via modification of sympathetic tone [13]. GPR55 is expressed in the brain, gut, and bone, and activation increases intracellular calcium concentrations via phospholipase C, RhoA, and G12 pathways [14]. GPR119 is expressed in pancreatic cells, and by increasing CamP production, insulin production also increases. Both GPR55 and GPR119 can also be activated by OEA and PEA [15]. AEA is also known to be a selective blocker of TASK-1, a potassium ion channel located in the brain and in the carotid body involved in hypoxia response and one of the culprits in the development of sleep apnea [16].

This is, of course, only a brief introduction to major players in the ECS. The rest of this book is dedicated to further describing their roles in nearly every homeostatic function of human physiology. Before moving on though, I would like to touch on the difficulties of actually measuring the ECS. In experimental animal models, tissues can be easily tested for the presence of endocannabinoids, although for reasons detailed above, time is of the essence! Quantitative measures of the number of cannabinoid receptors as well as degradation and synthesis enzymes present in tissue samples can also be achieved with Western blot analysis or immunofluorescence, and Northern blot can be used to quantify mRNA to assess active gene expression of ECS proteins [17]. However, in actual patients, these sorts of tissue samples are not always possible and rarely feasible. In at least one of the studies detailed in this book, PET scans were used to access CB1 receptor activity in patients. Another study used CSF from patients to assess endocannabinoid levels; however, this would only give a more global CNS picture and say little about levels in individual structures. Serum endocannabinoid levels are often also collected, once again with the limitation of giving only a very global assessment at only a single point in time. And, given the majority of the mechanism described above, which are typically autocrine and paracrine in nature, in addition to the considerable variation in tissues' ECS activity, this approach provides limited information as well. Endocannabinoid levels in neither hair nor urine correlated well with plasma levels [18]. Salivary ECS levels also do not appear to correlate to serum levels but did change significantly with acute stress [19]. Thus, lack of quick and reliable measurements in specific tissues of living patients therefore continues to be the primary obstacle in further diagnosing new diseases of the ECS. Hopefully, new techniques can be developed, as the next 14 chapters detail, the potential for ECS dysfunction and disease is myriad.

References

1. Murataeva N, Straiker A, Mackie K. Parsing the players: 2-arachidonoylglycerol synthesis and degradation in the CNS. Br J Pharmacol. 2014;171(6):1379–91.
2. Chicca A, Marazzi J, Nicolussi S, Gertsch J. Evidence for bidirectional endocannabinoid transport across cell membranes. J Biol Chem. 2012;287(41):34660–82. https://doi.org/10.1074/jbc.M112.373241.
3. Kaczocha M, Haj-Dahmane S. Mechanisms of endocannabinoid transport in the brain. Br J Pharmacol. 2022;179(17):4300–10. https://doi.org/10.1111/bph.15469.
4. Zou S, Kumar U. Cannabinoid receptors and the endocannabinoid system: signaling and function in the central nervous system. Int J Mol Sci. 2018;19(3):833. https://doi.org/10.3390/ijms19030833.
5. Fu J, Bottegoni G, Sasso O, Bertorelli R, Rocchia W, Masetti M, Guijarro A, Lodola A, Armirotti A, Garau G, Bandiera T, Reggiani A, Mor M, Cavalli A, Piomelli D. A catalytically silent FAAH-1 variant drives anandamide transport in neurons. Nat Neurosci. 2011;15(1):64–9. https://doi.org/10.1038/nn.2986.
6. Hillard CJ. Circulating endocannabinoids: from whence do they come and where are they going? Neuropsychopharmacology. 2018;43(1):155–72. https://doi.org/10.1038/npp.2017.130.
7. Mock ED, Gagestein B, van der Stelt M. Anandamide and other N-acylethanolamines: A class of signaling lipids with therapeutic opportunities. Prog Lipid Res. 2023;89:101194. https://doi.org/10.1016/j.plipres.2022.101194.
8. Wei F, Zhao L, Jing Y. Signaling molecules targeting cannabinoid receptors: Hemopressin and related peptides. Neuropeptides. 2020 Feb;79:101998. https://doi.org/10.1016/j.npep.2019.101998.
9. Hofer SC, Ralvenius WT, Gachet MS, Fritschy JM, Zeilhofer HU, Gertsch J. Localization and production of peptide endocannabinoids in the rodent CNS and adrenal medulla. Neuropharmacology. 2015;98:78–89. https://doi.org/10.1016/j.neuropharm.2015.03.021.
10. Petrucci V, Chicca A, Glasmacher S, Paloczi J, Cao Z, Pacher P, Gertsch J. Pepcan-12 (RVD-hemopressin) is a CB2 receptor positive allosteric modulator constitutively secreted by adrenals and in liver upon tissue damage. Sci Rep. 2017;7(1):9560.
11. Tominaga M, Tominaga T. Structure and function of TRPV1. Pflugers Arch. 2005;451(1):143–50. https://doi.org/10.1007/s00424-005-1457-8.
12. Montaigne D, Butruille L, Staels B. PPAR control of metabolism and cardiovascular functions. Nat Rev Cardiol. 2021;18(12):809–23. https://doi.org/10.1038/s41569-021-00569-6.

13. Penumarti A, Abdel-Rahman AA. The novel endocannabinoid receptor GPR18 is expressed in the rostral ventrolateral medulla and exerts tonic restraining influence on blood pressure. J Pharmacol Exp Ther. 2014;349(1):29–38. https://doi.org/10.1124/jpet.113.209213.
14. Lauckner JE, Jensen JB, Chen HY, Lu HC, Hille B, Mackie K. GPR55 is a cannabinoid receptor that increases intracellular calcium and inhibits M current. Proc Natl Acad Sci USA. 2008;105(7):2699–704. https://doi.org/10.1073/pnas.0711278105.
15. Im DS. GPR119 and GPR55 as receptors for fatty acid ethanolamides, oleoylethanolamide and palmitoylethanolamide. Int J Mol Sci. 2021;22(3):1034. https://doi.org/10.3390/ijms22031034.
16. Maingret F, Patel AJ, Lazdunski M, Honoré E. The endocannabinoid anandamide is a direct and selective blocker of the background K(+) channel TASK-1. EMBO J. 2001;20(1–2):47–54. https://doi.org/10.1093/emboj/20.1.47.
17. Wager-Miller J, Mackie K. Western blotting of the endocannabinoid system. Methods Mol Biol. 2023;2576:385–93. https://doi.org/10.1007/978-1-0716-2728-0_32.
18. Valdivieso Barba S, Kirschbaum C, Gao W. Endocannabinoid and perceived stress: association analysis of endocannabinoid levels in hair versus levels in plasma and urine. Biol Psychol. 2023;178:108541. https://doi.org/10.1016/j.biopsycho.2023.108541.
19. Ney L, Stone C, Nichols D, Felmingham K, Bruno R, Matthews A. Endocannabinoid reactivity to acute stress: investigation of the relationship between salivary and plasma levels. Biol Psychol. 2021;159:108022. https://doi.org/10.1016/j.biopsycho.2021.108022.

The Skin

2

The skin is not the first organ that comes to mind for most people when they think of the endocannabinoid system (ECS). Other than its outward appearance, most of us do not consider our skin much at all. It does, however, play a very important role of protecting all our other more attention-grabbing organs from the outside world. It thermoregulations the extremes of cold and heat as well as being one of first lines of defense against infection. We also rely upon it for our tactile sense of the world around us. And, like other tissues of the body, it is riddled with cannabinoid receptors. The ECS therefore plays an important role not only in the normal homeostatic function of our skin, but its dysfunction could also drive many diseases of the skin [1].

2.1 Distribution of ECS in the Skin

CB1 and CB2 receptors are present in keratinocytes, the flat cells that make up the majority of what we consider skin, as well as on sebaceous glands, hair follicles, melanocytes (responsible for skin color), and sweat glands. Keratinocytes and melanocytes can also produce both 2-AG and anandamide (AEA). Our skin is also richly interlaced with nerves and vascular tissues that are themselves contain large concentrations of cannabinoid receptors. Resident immune cells like dendritic cell, mast cells, T and B cells, and macrophages are all fully capable of interacting with the ECS too. TRP receptors and PPAR gamma receptors that can be activated by endocannabinoids are also present in the majority of skin cells. In short, the skin contains all the components for a fully functional ECS. How those components interact, the effects these interactions create are still an area of active investigation.

J.-P. H. Dedam et al., *Diseases of the Endocannabinoid System*,
https://doi.org/10.1007/978-3-031-87484-0_2

2.2 Role of ECS in Hair Growth

In our hair follicles, only CB1 is expressed and only during one portion of the hair growth cycle. This cycle begins with anagen, a rapid phase of growth; followed by catagen, when growth stops; and then telogen, a quiet period when the hair falls out and the follicle prepares to grow again. AEA plays a role in inducing catagen and slowing hair growth. Cannabis is likely therefore a very poor hair loss treatment! Hair growth is complex, though. Low levels of AEA are known to increase lipid production in sebaceous glands located in hair follicles, which is vital for maintaining healthy hair via CB2 receptor activation. Overproduction of lipids in sebaceous glands could also result in conditions such as acne [2]. Another driver of hair loss is mast cell activation and release of histamine. This can also be decreased by AEA via CB2 activation. An imbalance in the activation of CB1 versus CB2 receptors could therefore be a culprit in some types of hair loss.

2.3 Regulation of Immune System in the Skin

Another important role of the ECS in the skin is regulation of the immune response. AEA in the skin decreases pro-inflammatory cytokine effect on keratinocytes including those from TNF alpha and INF gamma and could help to prevent excessive responses by co-located T cells via CB1-dependent pathway [3]. Both AEA and 2-AG can increase macrophage release of anti-inflammatory factors like NO, IL 6, and PGE-2 via a CB2 receptor pathway. CB2-dependent blockage of leukocyte migration into tissues has also been shown in experimental models to reduce fibrosis [4]. Coupled with the previously mentioned mast cell deactivation, the ECS has a potent anti-inflammatory effect that has borne out in clinical studies as well. It is unknown at this point if the antimicrobial effects of cannabis itself is enough to prevent infection in endocannabinoid-activated and relatively immunosuppressed skin. Given the long history of topical cannabis preparations for inflammatory and infectious conditions, this seems very likely and not a clinical concern.

2.4 Regulation of Sensation in the Skin

Sensation in the skin is undoubtably intimately linked to the ECS. CB1, CB2, and TRPV receptors are all involved in both pain and itch. CB receptor activation has been shown in numerous studies to decrease itching. Both centrally and peripherally, itch has also been reduced by ECS decreases in mast cell release of histamine mentioned above. Endocannabinoids desensitize TRPV1 receptors similar to capsaicin also, thereby reducing itch. Pain will be covered in great detail in the nervous system section of this book [5].

2.5 Regulation of Angiogenesis in the Skin

The ECS is also known to be involved in angiogenesis. The growth of new blood vessels is essential for tissue growth, including cancer as well as wound healing. Both AEA and CB1 have been shown to block new capillary growth, potentially having a beneficial effect on development of cancer but limiting wound healing [6]. This topic is further discussed in Chap. 15, "Cell Death and Growth." What then, if any, is the role of the endocannabinoid system in dysregulated skin?

2.6 Atopic Dermatitis

Atopic dermatitis is a common inflammatory skin disorder that occurs when keratinocytes fail to develop into organized, flat layers. The constant renewal of our skin is dependent on the proliferation of keratinocytes at the lower levels of the epidermis, their differentiation as they move toward the surface into well-organized flat layers, and eventually the shedding of dead cells over the course of a 28-day period. Impaired differentiation and disrupted orderly layers allow allergens to penetrate through a normally impermeable layer and activate the immune system lurking beneath the surface causing inflamed, swollen, itchy skin. The ECS has a role in the development of keratinocytes as evidenced by CB receptor knockout mice. CB1 knockout mice had impaired development and differentiation of keratinocytes and were slower to repair impermeability, indicating CB1 importance in creating and maintaining a normal skin barrier. CB2 knockout mice demonstrated increased recovery of impermeability. The exact mechanism and therefore potential treatment strategies remain unknown [7]. While CB2 activation may not have positive effects on keratinocyte development, the combined effects of CB1 and CB2 are needed to prevent excessive allergic immune response in the skin, and increasing endocannabinoids had a beneficial effect on dermatitis in experimental models [8]. Pruritus, a significant symptom in atopic dermatitis and many other dermatologic conditions, can be improved with blockage of FAAH and MAGL, both systemically and at the level of the spinal cord, which would lead to physiological increases in endocannabinoids [9]. The disruption of keratinocyte differentiation and overactivation of the immune system are also hallmarks of psoriasis, although the exact mechanism for this disease process is still unknown. It is possible, therefore, that atopic dermatitis, contact dermatitis, and psoriasis are all at least in part the result of a dysfunctional ECS.

References

1. Río CD, Millán E, García V, Appendino G, DeMesa J, Muñoz E. The endocannabinoid system of the skin. A potential approach for the treatment of skin disorders. Biochem Pharmacol. 2018 Nov;157:122–133. https://doi.org/10.1016/j.bcp.2018.08.022. Epub 2018 Aug 20. PMID: 30138623.

2. Dobrosi N, Toth BI, Nagy G, Dozsa A, Geczy T, Nagy L, et al. Endocannabinoids enhance lipid synthesis and apoptosis of human sebocytes via cannabinoid receptor-2-mediated signaling. FASEB J. 2008;22(10):3685–95.
3. Chiurchiu V, Rapino C, Talamonti E, Leuti A, Lanuti M, Gueniche A, et al. Anandamide suppresses proinflammatory T cell responses in vitro through type-1 cannabinoid receptor-mediated mTOR inhibition in human keratinocytes. J Immunol. 2016;197(9):3545–53.
4. Akhmetshina A, Dees C, Busch N, Beer J, Sarter K, Zwerina J, Zimmer A, Distler O, Schett G, Distler JH. The cannabinoid receptor CB2 exerts antifibrotic effects in experimental dermal fibrosis. Arthritis Rheum. 2009;60(4):1129–36. https://doi.org/10.1002/art.24395.
5. Avila C, Massick S, Kaffenberger BH, Kwatra SG, Bechtel M. Cannabinoids for the treatment of chronic pruritus: a review. J Am Acad Dermatol. 2020;82(5):1205–12. https://doi.org/10.1016/j.jaad.2020.01.036.
6. Pisanti S, Borselli C, Oliviero O, Laezza C, Gazzerro P, Bifulco M. Antiangiogenic activity of the endocannabinoid anandamide: correlation to its tumor-suppressor efficacy. J Cell Physiol. 2007;211(2):495–503. https://doi.org/10.1002/jcp.20954.
7. Roelandt T, Heughebaert C, Bredif S, Giddelo C, Baudouin C, Msika P, Roseeuw D, Uchida Y, Elias PM, Hachem JP. Cannabinoid receptors 1 and 2 oppositely regulate epidermal permeability barrier status and differentiation. Exp Dermatol. 2012 Sep;21(9):688–93. https://doi.org/10.1111/j.1600-0625.2012.01561.x. PMID: 22897575.
8. Karsak M, Gaffal E, Date R, Wang-Eckhardt L, Rehnelt J, Petrosino S, et al. Attenuation of allergic contact dermatitis through the endocannabinoid system. Science. 2007;316(5830):1494–7.
9. Yesilyurt O, Cayirli M, Sakin YS, Seyrek M, Akar A, Dogrul A. Systemic and spinal administration of FAAH, MAGL inhibitors and dual FAAH/MAGL inhibitors produce antipruritic effect in mice. Arch Dermatol Res. 2016;308(5):335–45. https://doi.org/10.1007/s00403-016-1649-4.

The Gi Tract

3

3.1 Distribution of ECS in GI Tract

Our guts are often called our second brains due the enormous number of neurons present. Much like our brains, these neurons are packed with cannabinoid receptors. Aside from the nerve cells servicing our GI tracts, cannabinoid receptors are also present in gastric and intestinal lining cells, intestinal smooth muscle cells, adjacent immune cells, and blood vessels, and they are vital for the normal functioning of the GI tract.

3.2 Salivation

Digestion begins mostly with chewing and salivation. Cannabis users have long struggled with decreased salivation after cannabis use, a phenomenon often called "cotton mouth." CB1 receptors are present in cholinergic neurons innervating the submandibular glands, but not in the glands themselves. In experimental models, THC was capable of decreasing salivation, and this effect was reversed by CBD in a dose-dependent manner [1]. The role of the endocannabinoid system in salivation may be to decrease salivation in inflammatory states. Experiments showed that inhibition of salivation by lipopolysaccharides, substances produced by pathogenic bacteria that activate the immune system to drive inflammation, was partially prevented by administration of CB1 receptor blockers [2]. CB1 receptor activation in the hypothalamus can also decrease salivation by blocking parasympathetic stimulation of the submandibular glands [3].

J.-P. H. Dedam et al., *Diseases of the Endocannabinoid System*,
https://doi.org/10.1007/978-3-031-87484-0_3

3.3 GI Motility and IBS

Cannabis has been used for millennium as an anti-diarrheal treatment. It was not until 1998, however, when the mechanism behind this action could be understood. Cannabinoid receptor 1 was found presynaptically to human ileum smooth muscles cells, the third portion of the small intestine, and its activation resulted in inhibition of normal contraction by blocking the release of acetylcholine from the presynaptic neuron [4].

Cannabinoid receptors can slow GI smooth muscle contraction via a different pathway as well, the peristaltic reflex. Peristalsis involves complex coordination of sensory neurons, interneurons, and excitatory and inhibitory motor neurons as well as smooth muscles themselves. Food boluses stimulate the sensory neurons. Excitatory motor neurons upstream from the bolus release substance P and acetylcholine to cause smooth muscles to contract and propel the bolus on. Inhibitory motor neurons downstream from the bolus release nitric oxide and vasoactive intestinal polypeptide (VIP) to relax smooth muscles allowing the bolus to pass. Activation of CB1 receptors inhibits contractions from the pre-bolus section of the intestine by blocking presynaptic acetylcholine release but also the release of substance P in the same area. Downstream from the bolus, CB1 activation inhibits the release of VIP, although this is likely via an interneuron as the VIP-releasing neurons do not express a CB1 receptor. The end result is again the slowing of the normal peristaltic reflex [5]. Central administration of peptide cannabinoid or pepcan hemopressin resulted in slowing of GI transit time via CB1 activation in the brain, but not CB2 [6]. Other studies have also demonstrated a decrease in feeding behavior and analgesic effects in mice [7].

Alterations in the ECS have been associated with irritable bowel syndrome. Irritable bowel syndrome is a clinical diagnosis, meaning that there is no lab test or imaging study to determine if a patient has this condition. The diagnosis is made based on the presence of recurrent pain, at least once per week for at least 3 months associated with at least two of the following factors: pain with defecation, associated change in frequency of stool, or change in stool appearance. There are three subsets of this disease: constipation-predominant, diarrhea-predominant, or mixed. In IBS patients with predominate constipation, there were higher levels of FAAH in biopsy, which could result in the overall slower transit time with slower turnover in intestinal endocannabinoids [8]. Patients with a CB1 R polymorphism that affected functions showed a higher association with diarrhea-predominant IBS than healthy controls and, in a separate study, were shown to have faster than normal colonic transit time, which can be decreased by administration of dronabinol, a synthetic form of THC [9, 10].

Another receptor thought to be related to the endocannabinoid system, G protein-coupled receptor 55 or GPR55 is also found in large concentrations in the GI tract, specifically on neurons affecting motility in the colon. Currently, this receptor is still characterized as an orphan receptor, meaning we do not completely understand what normally activates it in the body, but we do know that it can be affected by plant and synthetic cannabinoids. In a study in mice, strips of the colon had reduced muscle contractions after being treated with a synthetic GPR55 agonist, and this effect was reversed by CBD. What clinical effect this has in humans is still not understood, although GPR55 is also known to be involved in metabolism and insulin release [11].

CB2 receptor has also been shown to have an effect on motility but only in disease states. Pathogenic bacteria in the gut can also produce lipopolysaccharides that bind receptors on immune cells causing them to release pro-inflammatory signals like TNF seen above, which in turn increase GI motility to expel these bacteria from the system. CB2 receptor activation blocks this mechanism [12]. While this might not seem like a beneficial effect, anyone who has ever taken an anti-diarrheal medication for traveler's diarrhea can tell you, it can be the difference between a slightly uncomfortable bus ride and disaster. In addition to slowing GI motility, CB1 activation has also been shown to decrease excess secretion intestines of cholera-infected mice. Excess secretion fluid and electrolytes in the intestine provoked by cholera toxin leads to severe diarrhea, eventually resulting in dehydration and death as well as spreading the disease in areas where there is not proper sanitation [13].

3.4 Visceral Sensitivity and IBS

Visceral sensitivity, or our sense of pain or discomfort in the GI tract, is also powerfully influenced by the ECS. Rat models have shown that activation of CB1 receptor in the colon reduced distention-related pain (Bruberg) [14]. Hypersensitivity to visceral stimulation is thought to be another mechanism underlying irritable bowel syndrome. Human biopsies of IBS patients have shown an increase in TRPV1 receptors, which mediate pain [15]. Chronic stress in rats increases TRPV1 expression and decreases CB1 receptor expression [16]. This is a likely mechanism for how chronic stress and an increase in pain sensitivity can lead to chronic pain in conditions like IBS, with an overabundance of pain-sensitizing receptors (TRPV1) and fewer pain-desensitizing receptors (CB1). In Korean populations of IBS sufferers, patients with IBS were found to have higher frequency of ATT repeating triplets in the CB1 receptor gene. The more repeats, the less well the receptor functioned and more severe the patients' symptoms [17]. Patients suffering from IBS often report relief in symptoms from cannabis use, increasing the likelihood that at least some subsets of IBS is caused by an endocannabinoid deficiency syndrome as described above.

3.5 Anti-inflammatory Effects of the ECS and IBD

The ECS has long been known to have a powerful anti-inflammatory effect on the immune system via CB2 receptor activation, and there is ample evidence to suggest that it has a role in inflammatory bowel Diseases like celiac disease, ulcerative colitis, and Crohn's disease. Ulcerative colitis and Crohn's disease together are called inflammatory bowel diseases or IBD (distinct from IBS discussed above). IBD is characterized by severe inflammation of the GI tract, just the colon in the case of ulcerative colitis and the entirety of the GI tract in Crohn's disease. The exact cause of IBD is not known, but both genetic and environmental conditions play a role in its development. A single nucleotide polymorphism in CB1 receptor has been shown in pediatric inflammatory bowel disease patients to increase the risk of developing disease as well as resulting in more severe diseases once diagnosed [18]. A single nucleotide polymorphism of the FAAH gene has been shown to lead to earlier disease onset in ulcerative colitis as well as increased fistula formation in Crohn's disease [19]. Single nucleotide polymorphism in the CB1 receptor can also increase risk of Ulcerative colitis and lower age of presentation in Crohn's [20]. Activation of CB2 receptor in human colon cells blocks TNF alpha driven production of Interleukin 8. This interleukin is known to attract neutrophils and is a major component of the pathology of inflammatory bowel disease [21]. ECS components CB1, AEA, and FAAH activities have all found to be increased in experimental intestinal inflammation models as well as from biopsy from patients with different inflammatory conditions including celiac disease and IBD and later decrease back to normal once remission is obtained. Blockage of CB1 in animal models causes inflammation to worsen, but inhibition of FAAH decreases that inflammation. Increased endocannabinoid tone protects against epithelial damage and increased motility during inflammation [22]. CB1 receptors present directly on colon cells have been shown to facilitate wound healing when stimulated by AEA in animal models, a process that is significantly delayed in IBD [23]. As with IBS, many patients suffering from IBD use cannabis as a way of controlling their symptoms, which include pain, cramping, and diarrhea. While the mechanism of these benefits is obvious based on the role of the ECS in the gut, whether cannabis can have a more powerful and direct effect on the progression of IBD requires further study. Chronic inflammation from IBD is also associated with higher risk of colorectal cancer. Analysis of cannabinoid receptors in colon cancer tumors showed that they had fewer CB1 receptors than normal colon tissues but significantly more CB2. In vitro activation of both receptors in the tumor cells resulted in apotheosis, or programed cell death of the cell [24]. The role of the ECS and tumor suppression will be addressed in further detail in another chapter.

3.6 Celiac Disease

Celiac disease is another chronic inflammatory disease affecting the GI tract. Here, consumption of gluten, a protein found in wheat, causes activation of the immune system in the small intestine. This inflammation results in significant atrophy of small intestine's villi where the majority of nutrient absorption takes places. The result is pain, bloating, diarrhea, and malabsorption. There is a significant genetic predisposition to celiac disease that is not completely understood. Genes that encode proteins called HLA that allow the T cells in our immune systems to interact with pathogens are involved but not in all cases. A single nucleotide polymorphism found in CB2 receptor in children in Southern Italy resulted in more than sixfold increase in likelihood of developing celiac disease [25]. Being of Southern Italian descent myself, I can think of fewer cruel twists of fate than large numbers of children there not being able to eat pasta. The study also showed increased expression of CB2 receptors in small intestine biopsies of affected children with massive infiltration of inflammatory T cells. Previous studies have shown that AEA, via CB2, caused suppression but not death of these cells as well as decreasing the release of IL-17, a pro-inflammatory cytokine implicated in celiac disease pathogenesis [26]. Without a properly functioning CB2 receptor, these patients are likely unable to control an out-of-control immune response to gluten.

3.7 Gut Permeability

Gut permeability is another significant component of inflammatory bowel disease and other inflammatory conditions. This allows bacteria and other inflammatory elements to pass from the inside of the bowel into the tissues themselves where they will activate our immune system. Permeability is controlled by a number of different proteins that hold the epithelial cells together and is increased with inflammation. In a study using human intestinal cells in vitro treated first with inflammatory cytokines TNF and interferon gamma and endogenous cannabinoid, AEA and 2-AG administration increased permeability, while administration of THC and CBD decreased permeability. Both effects were blocked by CB1 receptor antagonists, indicating that the effect is via this receptor and not CB2 [27]. Why THC improves permeability and AEA and 2-AG worsen it is not understood. Another study looking at inflammation induced in human colon cells by hypoxia showed that TRPV1and PPAR gamma activation resulted in decreased permeability as well as two other ligands known to interact with endocannabinoid receptors: OEA and PEA. Their presence also had effects on permeability. The exact mechanism is likely quite complex and depends on multiple receptors and different ligands in an entourage-like effect [28].

3.8 ECS Control of Gastric Acid and Ulcer Prevention

In the stomach, CB1 receptor activation in rat models resulted in delayed gastric emptying. However, CB1 antagonist did not have any physiological effect, indicating that there appears to be no tonic activity of the ECS on gastric motility as there is thought to be in the intestines [29]. CB1 receptor activation in the stomach also results in decreased acid secretion, although the exact mechanism is not known but is most like via peripheral vagus neurons [30]. This effect is significant enough to prevent ulcer formation in rats that are water stressed [31]. CB1 activation by THC has also been shown to prevent gastric ulcer and bleeding due to NSAIDs in mouse models at doses below levels that would induce neuro-psychologic side effects [32]. These models showed that in addition to decreased gastric acid secretion, there was also a decrease in pro-inflammatory cytokines such as IL-1 and TNF directly related to ECS activation. Enteroendocrine cells such as cells producing CCK, insulin, and glucagon-like peptide 1 also appear to express components of the ECS potentially affecting nutrient intake and metabolism. The ECS tone is significantly increased in obesity. This will be covered in greater detail in the chapter on metabolism. While CB1 was cited to prevent stomach ulcers in mice models, activation of CB2 also appears to have a similar gastroprotective effect by reducing elevated TNF-alpha and IL-1 levels in alcohol and NSAIDs mice models [33]. In a study in mice, centrally mediated angiotensin II gastroprotective effects prevented alcohol-induced ulceration in wild type but not CB1 knockout mice [34]. This finding raises the possibility that an endocannabinoid deficiency could result in increased risk of gastric ulcers.

3.9 ECS and Bowel Flora

We cannot talk about our GI tract and not talk about the bacteria that reside there as well. The number and type of bacteria living there are of vital importance to normal physiology as well as pathology. Our ECS has been shown in numerous studies to be reactive to these bacteria. Exposure of human colon cells in vitro to *Lactobacillus* induced expression of CB2 receptor. In the same study, administration of *Lactobacillus* in mice resulted in decreased pain sensitivity to dilation, which was reversed by a CB2 receptor antagonist, indicating that CB2 receptor has a role in pain as well (Rosseaux) [35]. Mice raised without any gut bacteria are known to have significant differences in ECS components and are partially restored by fecal transplant [36]. While the effect of bacteria on ECS is likely complex and requires a great deal of study and attention, a recent study also showed that the ECS can have direct effects on pathogenic bacteria. Mice with elevated 2-AG levels were protected from *Enterobacteriaceae* pathogens, which are also common human pathogens. This effect was by direct antagonization of a receptor in the bacteria needed to secrete pathogenic chemicals by 2-AG, thus preventing inflammation and diarrhea [37].

3.10 ECS and Control of Nausea and Vomiting

At the being of this chapter, I referred to the gut as the second brain. The connection between these two brains is the vagus nerve, a bidirectional signal pathway vital for normal physiologic function. Vagus neurons express CB1 receptors and their activation influences many different physiologic processes. Studies in CB1 receptor knockout mice have shown its necessity in controlling normal GI motility [38]. Cannabinoids' anti-nausea properties are also medicated via the vagus nerve. Nausea and vomiting are bodily defense mechanisms to expel potentially dangerous substances from the body. However, in a variety of disease states, normal physiological changes like pregnancy and treatments like chemotherapy can trigger this response out of proportion to what is needed. Most instances of nausea and vomiting are triggered when serotonin is released from GI endothelial cells, leading to the activation of afferent vagus nerve fibers connecting to the brain. This then activates the dorsal vagal complex in the brain stem, which initiates the vomiting reflex. Activation of CB1 receptors present in this part of the brain can suppress this action. Blockage of CB1 receptor in the brain actually results in severe nausea and vomiting [39]. An interesting study of people who experienced motion sickness with parabolic flight maneuvers showed that they had lower levels of anandamide than those without symptoms. Following flight maneuvers, the subjects who did not experience motion sickness had higher than normal levels of anandamide, indicating not only its protective physiologic role in preventing motion sickness but making motion sickness another likely endocannabinoid deficiency syndrome [40]. Dronabinol is FDA approved for the treatment of chemotherapy-induced nausea as a second-line treatment. It also has the additional benefit of stimulated appetite and enjoyment of food, which is often a significant issue in cancer and cancer treatment. Endocannabinoids present in the developing brain and human breast milk are vital for infant suckling response [41]. In adults, CB1 receptor activation at multiple brain areas is involved in appetite and food intake. Activation in the nucleus accumbens, a key part of motivation and addiction pathways, stimulates both the desire to eat and pleasure in eating. 2-AG levels in the hypothalamus rise in a fasting state, encouraging eating, and then fall again after eating [42]. A CB1 receptor antagonist, rimonabant, actually made it to clinical trials as a weight loss drug. However, severe nausea and vomiting as well increased depression symptoms and suicide attempts prevented it from reaching the market despite it being very effective at weight loss.

3.11 Cyclical Vomiting Syndrome and Cannabis Hyperemesis Syndrome

Perhaps the highest profile conditions potentially involving the ECS and the GI tract are cyclic vomiting syndrome (CVS) and cannabinoid hyperemesis syndrome (CHS) as both conditions are on the rise, and these conditions appear to be related to the increase in high potency cannabis products. The exact underlying mechanism of both syndromes remains unknown, so both are defined by Rome IV clinical criteria. For cyclic vomiting syndrome, there are three criteria: stereotypical episodes of vomiting, at least three episodes per year, and absence of vomiting in-between. For cannabinoid hyperemesis syndrome, the criteria are the same as the addition of prolonged, high-dose THC consumption and cessation of symptoms with sustained abstinence [43]. Because of cannabis's known antiemetic effects noted above, a high percentage of cyclic vomiting patients are known to use cannabis, making the potential of blurring the lines between the two syndromes high. CVS and CHS also share other supporting characteristics such as relief obtained by hot showers and similar comorbidity profiles. The best criterion for differentiation of the two in one study was the complete and persistent resolution of symptoms with cessation of cannabis [44]. CVS has been noted in a number of human studies to have alterations in the ECS, such as increased plasma levels of circulating 2-AG and OEA [45]. Polymorphic genetic analysis of CB1 gene seems to confirm this with CNR1 rs806368 AG alleles imparting increased risk of CVS. However, CNR1 rs806368 CC allele decreased the risk of CVS [46]. The first polymorphism, the AG allele, has been shown in other analysis to increase ECS tone and central obesity, an association to be discussed in the following chapter on metabolism [47]. The second polymorphism, the CC allele, likely decreases ECS and is linked to increased panic attack risk in a separate study [48]. It would appear from this simple analysis that patients with CVS may suffer from a relative ECS excess, despite its beneficial effects on nausea and vomiting. In all likelihood, the underlaying mechanism, which is still unknown, is more complicated than a simple excess syndrome. The dysregulation of the endocannabinoid system apparent in CVS is also certainly occurring in CHS. Chronic and high-dose consumption of THC has long been known to have significant effects on the ECS. Rat Brian studies showed significant downregulation of CB1 receptor after repeated exposure to THC [49]. This process results in tolerance and later potential withdrawal symptoms after cessation of THC, characterized by loss of appetite and even stomach pain [50]. Given relief with hot showers, it is not surprising that polymorphisms in the TRPV1 gene have been found in CHS sufferers, as this is one of the receptors activated in the body in exposure to heat and can also be activated by THC as well as endocannabinoids. Also implicated were dopamine receptors and CYP 450 enzymes that break down exogenous cannabinoids such as THC [51]. As more is learned about the underlaying mechanism of these two syndromes, a better understanding of how and where the ECS is dysfunctional will likely come to light.

3.12 ECS and Liver Disease

Under healthy conditions, the endocannabinoid system is not particularly active in the liver, to the point that it was only recently that its role has been appreciated at all. CB1 receptors are, however, present in hepatocytes, stellate cells, myofibroblasts, and Kuffler cells, and their activity increased under pathologic conditions. CB1 receptor activation on hepatocytes promotes lipogenesis, increases insulin resistance, and increases gluconeogenesis. ECS tone is increased by high-fat diets and then in turn drives further fatty liver disease. CB1 receptors on hepatic stellate cells are activated by pro-fibrotic signals, which lead to further fibrosis. Blockage of CB1 receptor in the liver can prevent fibrosis [52]. Alcoholic fatty liver disease could also be reduced by blockage of CB1 receptors in mice models. Here, once again, it was shown that alcohol consumption leads to increased CB1 receptor expression and endocannabinoid tone, leading to further fatty infiltration, representing a significant exception to the usual function of the ECS as a feedback inhibitor [53]. In liver cirrhosis. CB1 receptor is unregulated in vascular cells within the liver and its activation causes vasodilation. This exacerbates with vasodilation normally seen in liver diseases, leading to further liver congestion, low blood pressures in rat models, and worsened ascites and portal hypertension, the major sources of morbidity and mortality in liver diseases [54].

CB2 receptors, on the other hand, are not present in hepatocytes but can be found in hepatocytes, stellate cells, myofibroblasts, and Kuffler cells as well as immune cells located within the liver. CB2 receptors are also unregulated in liver cirrhosis; however, they exert a powerful anti-fibrogenic effects, preventing progression and even reversing cirrhosis (Julien 2004). This was shown in mouse models to be accomplished by blocking the production of IL-17, a pro-inflammatory and pro-fibrotic signal produced by TH-17 lymphocytes [55]. While CB1 and CB2 have opposing effects in the body, chronic cannabis use is still a risk factor for progression of liver disease in hepatitis C, indicating a more dominant role of CB1 [56]. Selective activation of CB2, however, does represent a novel potential pharmacologic target for the treatment of liver diseases.

The ECS is involved from the first bite you take to the final step of elimination. Its normal function is absolutely vital and as we learn more about its dysfunction and the diseases it causes, it will only become more and more important as a therapeutic target.

References

1. Andreis K, Billingsley J, Naimi Shirazi K, Wager-Miller J, Johnson C, Bradshaw H, Straiker A. Cannabinoid CB1 receptors regulate salivation. Sci Rep. 2022;12(1):14182. https://doi.org/10.1038/s41598-022-17987-2.
2. Fernandez-Solari J, Prestifilippo JP, Ossola CA, Rettori V, Elverdin JC. Participation of the endocannabinoid system in lipopolysaccharide-induced inhibition of salivary secretion. Arch Oral Biol. 2010 Aug;55(8):583–90. https://doi.org/10.1016/j.archoralbio.2010.05.006. Epub 2010 Jun 9. PMID: 20542488.
3. Prestifilippo JP, Medina VA, Mohn CE, Rodriguez PA, Elverdin JC, Fernandez-Solari J. Endocannabinoids mediate hyposalivation induced by inflammogens in the submandibular glands and hypothalamus. Arch Oral Biol. 2013;58(9):1251–9. https://doi.org/10.1016/j.archoralbio.2013.04.003.
4. Croci T, Manara L, Aureggi G, Guagnini F, Rinaldi-Carmona M, Maffrand JP, Le Fur G, Mukenge S, Ferla G. In vitro functional evidence of neuronal cannabinoid CB1 receptors in human ileum. Br J Pharmacol. 1998;125(7):1393–5. https://doi.org/10.1038/sj.bjp.0702190.
5. Grider JR, Mahavadi S, Li Y, Qiao LY, Kuemmerle JF, Murthy KS, Martin BR. Modulation of motor and sensory pathways of the peristaltic reflex by cannabinoids. Am J Physiol Gastrointest Liver Physiol. 2009;297(3):G539–49. https://doi.org/10.1152/ajpgi.00064.2009.
6. Li XH, Lin ML, Wang ZL, Wang P, Tang HH, Lin YY, Li N, Fang Q, Wang R. Central administrations of hemopressin and related peptides inhibit gastrointestinal motility in mice. Neurogastroenterol Motil. 2016;28(6):891–9. https://doi.org/10.1111/nmo.12789.
7. Macedonio G, Stefanucci A, Maccallini C, Mirzaie S, Novellino E, Mollica A. Hemopressin peptides as modulators of the endocannabinoid system and their potential applications as therapeutic tools. Protein Pept Lett. 2016;23(12):1045–51. https://doi.org/10.2174/0929866523666161007152435.
8. Fichna J, Wood JT, Papanastasiou M, Vadivel SK, Oprocha P, Sałaga M, Sobczak M, Mokrowiecka A, Cygankiewicz AI, Zakrzewski PK, Małecka-Panas E, Krajewska WM, Kościelniak P, Makriyannis A, Storr MA. Endocannabinoid and cannabinoid-like fatty acid amide levels correlate with pain-related symptoms in patients with IBS-D and IBS-C: a pilot study. PLoS One. 2013;8(12):e85073. https://doi.org/10.1371/journal.pone.0085073.
9. Camilleri M, Kolar GJ, Vazquez-Roque MI, Carlson P, Burton DD, Zinsmeister AR. Cannabinoid receptor 1 gene and irritable bowel syndrome: phenotype and quantitative traits. Am J Physiol Gastrointest Liver Physiol. 2013;304(5):G553–60. https://doi.org/10.1152/ajpgi.00376.2012.

10. Wong BS, Camilleri M, Eckert D, Carlson P, Ryks M, Burton D, Zinsmeister AR. Randomized pharmacodynamic and pharmacogenetic trial of dronabinol effects on colon transit in irritable bowel syndrome-diarrhea. Neurogastroenterol Motil. 2012;24:358–65.
11. Li K, Fichna J, Schicho R, Saur D, Bashashati M, Mackie K, et al. A role for O-1602 and G protein-coupled receptor GPR55 in the control of colonic motility in mice. Neuropharmacology. 2013;71:255–63.
12. Duncan M, Mouihate A, Mackie K, Keenan CM, Buckley NE, Davison JS, Patel KD, Pittman QJ, Sharkey KA. Cannabinoid CB2 receptors in the enteric nervous system modulate gastrointestinal contractility in lipopolysaccharide-treated rats. Am J Physiol Gastrointest Liver Physiol. 2008;295(1):G78–87. https://doi.org/10.1152/ajpgi.90285.2008.
13. Izzo AA, Capasso F, Costagliola A, Bisogno T, Marsicano G, Ligresti A, Matias I, Capasso R, Pinto L, Borrelli F, Cecio A, Lutz B, Mascolo N, Di Marzo V. An endogenous cannabinoid tone attenuates cholera toxin-induced fluid accumulation in mice. Gastroenterology. 2003;125(3):765–74. https://doi.org/10.1016/s0016-5085(03)00892-8.
14. Brusberg M, Arvidsson S, Kang D, Larsson H, Lindström E, Martinez V. CB1 receptors mediate the analgesic effects of cannabinoids on colorectal distension-induced visceral pain in rodents. J Neurosci. 2009;29(5):1554–64. https://doi.org/10.1523/JNEUROSCI.5166-08.2009.
15. van Wanrooij SJ, Wouters MM, Van Oudenhove L, Vanbrabant W, Mondelaers S, Kollmann P, Kreutz F, Schemann M, Boeckxstaens GE. Sensitivity testing in irritable bowel syndrome with rectal capsaicin stimulations: role of TRPV1 upregulation and sensitization in visceral hypersensitivity? Am J Gastroenterol. 2014;109(1):99–109. https://doi.org/10.1038/ajg.2013.371.
16. Hong S, Zheng G, Wiley JW. Epigenetic regulation of genes that modulate chronic stress-induced visceral pain in the peripheral nervous system. Gastroenterology. 2015;148(1):148–57. https://doi.org/10.1053/j.gastro.2014.09.032.
17. Park JM, Choi MG, Cho YK, Lee IS, Kim SW, Choi KY, Chung IS. Cannabinoid receptor 1 gene polymorphism and irritable bowel syndrome in the Korean population: a hypothesis-generating study. J Clin Gastroenterol. 2011;45(1):45–9. https://doi.org/10.1097/MCG.0b013e3181dd1573.
18. Strisciuglio C, Bellini G, Miele E, Martinelli M, Cenni S, Tortora C, Tolone C, Miraglia Del Giudice E, Rossi F. Cannabinoid receptor 2 functional variant contributes to the risk for pediatric inflammatory bowel disease. J Clin Gastroenterol. 2018;52(5):37–43. https://doi.org/10.1097/MCG.0000000000000755.
19. Storr M, Emmerdinger D, Diegelmann J, Yüce B, Pfennig S, Ochsenkühn T, Göke B, Lohse P, Brand S. The role of fatty acid hydrolase gene variants in inflammatory bowel disease. Aliment Pharmacol Ther. 2009;29(5):542–51. https://doi.org/10.1111/j.1365-2036.2008.03910.x.
20. Storr M, Emmerdinger D, Diegelmann J, Pfennig S, Ochsenkühn T, Göke B, Lohse P, Brand S. The cannabinoid 1 receptor (CNR1) 1359 G/A polymorphism modulates susceptibility to ulcerative colitis and the phenotype in Crohn's disease. PLoS One. 2010;5(2):e9453. https://doi.org/10.1371/journal.pone.0009453.
21. Ihenetu K, Molleman A, Parsons ME, Whelan CJ. Inhibition of interleukin-8 release in the human colonic epithelial cell line HT-29 by cannabinoids. Eur J Pharmacol. 2003;458(1–2):207–15. https://doi.org/10.1016/s0014-2999(02)02698-5.
22. Di Marzo V, Izzo AA. Endocannabinoid overactivity and intestinal inflammation. Gut. 2006;55(10):1373–6. https://doi.org/10.1136/gut.2005.090472.
23. Wright K, Rooney N, Feeney M, Tate J, Robertson D, Welham M, Ward S. Differential expression of cannabinoid receptors in the human colon: cannabinoids promote epithelial wound healing. Gastroenterology. 2005;129(2):437–53. https://doi.org/10.1016/j.gastro.2005.05.026.
24. Cianchi F, Papucci L, Schiavone N, Lulli M, Magnelli L, Vinci MC, Messerini L, Manera C, Ronconi E, Romagnani P, Donnini M, Perigli G, Trallori G, Tanganelli E, Capaccioli S, Masini E. Cannabinoid receptor activation induces apoptosis through tumor necrosis factor alpha-mediated ceramide de novo synthesis in colon cancer cells. Clin Cancer Res. 2008 Dec 1;14(23):7691–700. https://doi.org/10.1158/1078-0432.CCR-08-0799. PMID: 19047095.
25. Rossi F, Bellini G, Tolone C, Luongo L, Mancusi S, Papparella A, Sturgeon C, Fasano A, Nobili B, Perrone L, Maione S, del Giudice EM. The cannabinoid receptor type 2 Q63R variant increases the risk of celiac disease: implication for a novel molecular biomarker and future therapeutic intervention. Pharmacol Res. 2012;66(1):88–94. https://doi.org/10.1016/j.phrs.2012.03.011.

26. Cencioni MT, Chiurchiù V, Catanzaro G, Borsellino G, Bernardi G, Battistini L, Maccarrone M. Anandamide suppresses proliferation and cytokine release from primary human T-lymphocytes mainly via CB2 receptors. PLoS One. 2010;5(1):e8688. https://doi.org/10.1371/journal.pone.0008688.
27. Alhamoruni A, Wright KL, Larvin M, O'Sullivan SE. Cannabinoids mediate opposing effects on inflammation-induced intestinal permeability. Br J Pharmacol. 2012;165(8):2598–610. https://doi.org/10.1111/j.1476-5381.2011.01589.x.
28. Karwad MA, Couch DG, Wright KL, Tufarelli C, Larvin M, Lund J, O'Sullivan SE. Endocannabinoids and endocannabinoid-like compounds modulate hypoxia-induced permeability in CaCo-2 cells via CB1, TRPV1, and PPARα. Biochem Pharmacol. 2019;168:465–72. https://doi.org/10.1016/j.bcp.2019.07.017.
29. Izzo AA, Mascolo N, Capasso R, Germanò MP, De Pasquale R, Capasso F. Inhibitory effect of cannabinoid agonists on gastric emptying in the rat. Naunyn Schmiedeberg's Arch Pharmacol. 1999;360(2):221–3. https://doi.org/10.1007/s002109900054.
30. Adami M, Frati P, Bertini S, Kulkarni-Narla A, Brown DR, de Caro G, Coruzzi G, Soldani G. Gastric antisecretory role and immunohistochemical localization of cannabinoid receptors in the rat stomach. Br J Pharmacol. 2002;135(7):1598–606. https://doi.org/10.1038/sj.bjp.0704625.
31. Elgohary R, Abdelsalam RM, Abdel-Salam OM, Khattab MM, Salem NA, El-Khyat ZA, Morsy FA. Protective effect of cannabinoids on gastric mucosal lesions induced by water immersion restrain stress in rats. Iran J Basic Med Sci. 2021;24(9):1182–9. https://doi.org/10.22038/ijbms.2021.54338.12213.
32. Kinsey SG, Cole EC. Acute Δ(9)-tetrahydrocannabinol blocks gastric hemorrhages induced by the nonsteroidal anti-inflammatory drug diclofenac sodium in mice. Eur J Pharmacol. 2013;715(1–3):111–6. https://doi.org/10.1016/j.ejphar.2013.06.001.
33. Salaga M, Zatorski H, Zielińska M, Mosinska P, Timmermans JP, Kordek R, Storr M, Fichna J. Highly selective CB2 receptor agonist A836339 has gastroprotective effect on experimentally induced gastric ulcers in mice. Naunyn Schmiedeberg's Arch Pharmacol. 2017;390(10):1015–27. https://doi.org/10.1007/s00210-017-1402-3.
34. Gyires K, Rónai AZ, Zádori ZS, Tóth VE, Németh J, Szekeres M, Hunyady L. Angiotensin II-induced activation of central AT1 receptors exerts endocannabinoid-mediated gastroprotective effect in rats. Mol Cell Endocrinol. 2014;382(2):971–8. https://doi.org/10.1016/j.mce.2013.10.002.
35. Rousseaux C, Thuru X, Gelot A, Barnich N, Neut C, Dubuquoy L, Dubuquoy C, Merour E, Geboes K, Chamaillard M, Ouwehand A, Leyer G, Carcano D, Colombel JF, Ardid D, Desreumaux P. Lactobacillus acidophilus modulates intestinal pain and induces opioid and cannabinoid receptors. Nat Med. 2007;13(1):35–7. https://doi.org/10.1038/nm1521.
36. Manca C, Boubertakh B, Leblanc N, Deschênes T, Lacroix S, Martin C, Houde A, Veilleux A, Flamand N, Muccioli GG, Raymond F, Cani PD, Di Marzo V, Silvestri C. Germ-free mice exhibit profound gut microbiota-dependent alterations of intestinal endocannabinoidome signaling. J Lipid Res. 2020;61(1):70–85. https://doi.org/10.1194/jlr.RA119000424.
37. Ellermann M, Pacheco AR, Jimenez AG, Russell RM, Cuesta S, Kumar A, Zhu W, Vale G, Martin SA, Raj P, McDonald JG, Winter SE, Sperandio V. Endocannabinoids inhibit the induction of virulence in enteric pathogens. Cell. 2020;183(3):650–65. https://doi.org/10.1016/j.cell.2020.09.022.
38. Vianna CR, Donato J Jr, Rossi J, Scott M, Economides K, Gautron L, Pierpont S, Elias CF, Elmquist JK. Cannabinoid receptor 1 in the vagus nerve is dispensable for body weight homeostasis but required for normal gastrointestinal motility. J Neurosci. 2012;32(30):10331–7. https://doi.org/10.1523/JNEUROSCI.4507-11.2012.
39. Malik Z, Baik D, Schey R. The role of cannabinoids in regulation of nausea and vomiting, and visceral pain. Curr Gastroenterol Rep. 2015;17(2):429. https://doi.org/10.1007/s11894-015-0429-1.
40. Choukèr A, Kaufmann I, Kreth S, Hauer D, Feuerecker M, Thieme D, Vogeser M, Thiel M, Schelling G. Motion sickness, stress and the endocannabinoid system. PLoS One. 2010;5(5):e10752. https://doi.org/10.1371/journal.pone.0010752.

41. Fride E, Bregman T, Kirkham TC. Endocannabinoids and food intake: newborn suckling and appetite regulation in adulthood. Exp Biol Med (Maywood). 2005;230(4):225–34. https://doi.org/10.1177/153537020523000401.
42. Kirkham TC, Williams CM, Fezza F, Di Marzo V. Endocannabinoid levels in rat limbic forebrain and hypothalamus in relation to fasting, feeding and satiation: stimulation of eating by 2-arachidonoyl glycerol. Br J Pharmacol. 2002;136:550–7.
43. Venkatesan T, Levinthal DJ, Li BUK, Tarbell SE, Adams KA, Issenman RM, Sarosiek I, Jaradeh SS, Sharaf RN, Sultan S, Stave CD, Monte AA, Hasler WL. Role of chronic cannabis use: cyclic vomiting syndrome vs cannabinoid hyperemesis syndrome. Neurogastroenterol Motil. 2019;31(Suppl 2):e13606. https://doi.org/10.1111/nmo.13606.
44. Blumentrath CG, Dohrmann B, Ewald N. Cannabinoid hyperemesis and the cyclic vomiting syndrome in adults: recognition, diagnosis, acute and long-term treatment. Ger Med Sci. 2017;15:Doc06. https://doi.org/10.3205/000247.
45. Bashashati M, Bradshaw HB, Johnson CT, Zuckerman MJ, Sarosiek J, McCallum RW, Sarosiek I. Plasma endocannabinoids and cannabimimetic fatty acid derivatives are altered in cyclic vomiting syndrome: the effects of sham feeding. J Investig Med. 2023;71(8):821–9. https://doi.org/10.1177/10815589231196591.
46. Wasilewski A, Lewandowska U, Mosinska P, Watala C, Storr M, Fichna J, Venkatesan T. Cannabinoid receptor type 1 and mu-opioid receptor polymorphisms are associated with cyclic vomiting syndrome. Am J Gastroenterol. 2017;112(6):933–9. https://doi.org/10.1038/ajg.2017.73.
47. Jaeger JP, Mattevi VS, Callegari-Jacques SM, Hutz MH. Cannabinoid type-1 receptor gene polymorphisms are associated with central obesity in a southern Brazilian population. Dis Markers. 2008;25(1):67–74. https://doi.org/10.1155/2008/841490.
48. Peiró AM, García-Gutiérrez MS, Planelles B, Femenía T, Mingote C, Jiménez-Treviño L, Martínez-Barrondo S, García-Portilla MP, Saiz PA, Bobes J, Manzanares J. Association of cannabinoid receptor genes (*CNR1 and CNR2*) polymorphisms and panic disorder. Anxiety Stress Coping. 2020;33(3):256–65. https://doi.org/10.1080/10615806.2020.1732358.
49. Romero J, Berrendero F, Manzanares J, Pérez A, Corchero J, Fuentes JA, Fernández-Ruiz JJ, Ramos JA. Time-course of the cannabinoid receptor down-regulation in the adult rat brain caused by repeated exposure to delta9-tetrahydrocannabinol. Synapse. 1998;30(3):298–308.
50. Connor JP, Stjepanović D, Budney AJ, Le Foll B, Hall WD. Clinical management of cannabis withdrawal. Addiction. 2022;117(7):2075–95. https://doi.org/10.1111/add.15743.
51. Russo EB, Spooner C, May L, Leslie R, Whiteley VL. Cannabinoid hyperemesis syndrome survey and genomic investigation. Cannabis Cannabinoid Res. 2022;7(3):336–44. https://doi.org/10.1089/can.2021.0046.
52. Maccarrone M, Bab I, Bíró T, Cabral GA, Dey SK, Di Marzo V, Konje JC, Kunos G, Mechoulam R, Pacher P, Sharkey KA, Zimmer A. Endocannabinoid signaling at the periphery: 50 years after THC. Trends Pharmacol Sci. 2015;36(5):277–96. https://doi.org/10.1016/j.tips.2015.02.008.
53. Jeong WI, Osei-Hyiaman D, Park O, Liu J, Bátkai S, Mukhopadhyay P, Horiguchi N, Harvey-White J, Marsicano G, Lutz B, Gao B, Kunos G. Paracrine activation of hepatic CB1 receptors by stellate cell-derived endocannabinoids mediates alcoholic fatty liver. Cell Metab. 2008;7(3):227–35. https://doi.org/10.1016/j.cmet.2007.12.007.
54. Bátkai S, Járai Z, Wagner JA, Goparaju SK, Varga K, Liu J, Wang L, Mirshahi F, Khanolkar AD, Makriyannis A, Urbaschek R, Garcia N Jr, Sanyal AJ, Kunos G. Endocannabinoids acting at vascular CB1 receptors mediate the vasodilated state in advanced liver cirrhosis. Nat Med. 2001 Jul;7(7):827–32. https://doi.org/10.1038/89953. PMID: 11433348.
55. Guillot A, Hamdaoui N, Bizy A, Zoltani K, Souktani R, Zafrani ES, Mallat A, Lotersztajn S, Lafdil F. Cannabinoid receptor 2 counteracts interleukin-17-induced immune and fibrogenic responses in mouse liver. Hepatology. 2014 Jan;59(1):296–306. https://doi.org/10.1002/hep.26598. Epub 2013 Nov 19. PMID: 23813495.
56. Mallat A, Hezode C, Lotersztajn S. Environmental factors as disease accelerators during chronic hepatitis C. J Hepatol. 2008;48(4):657–65. https://doi.org/10.1016/j.jhep.2008.01.004.

Metabolism

4

Modern cannabis use has long been associated with "the Munchies." In chapter on the GI track, we saw how this phenomenon is triggered by activation of CB1 receptors in the brain. What the body subsequently does with those calories is the subject of this chapter, and it should come as no surprise at this point that the ECS has a big hand to play, especially in fat accumulation. In addition to CB1 receptors in the hypothalamus, the central control of metabolism, they are also present peripherally in hepatocytes, skeletal muscles, adipose tissues, and the endocrine cells of the pancreas. We will explore how obesity, one of the most common diseases in western society, may be our first example of an endocannabinoid excess syndrome.

4.1 Central Control and Obesity

The hypothalamus, located in the brain, is the site where numerous signals from the body indicating nutritional state are received, integrated, and generated to regulate food intake to maintain energy balance. Numerous cannabinoid receptors are present there and control a complex cross talk of numerous signals including leptin, ghrelin, corticosteroids, orexin, neuropeptide Y, and alpha melanocyte-stimulating hormone. Leptin is a hormone produced in the body, often in adipose cells, that tells the hypothalamus when enough food has been consumed. It causes appetite-suppressing hormones (alpha-melanocyte-stimulating hormone) in the hypothalamus to be increased and appetite-stimulating hormones (neuropeptide Y) to be reduced. Mutations in the leptin receptor lead to significant obesity as the body is unable to turn off appetite stimulation. Mice studied with this condition have elevated endocannabinoids in the hypothalamus, but not the rest of the brain. Blocking CB1 receptors in these mice reduced appetite and restored normal leptin function, subsequently decreasing endocannabinoid levels in the hypothalamus to normal levels. All of this taken together would indicate a tonically active ECS in the hypothalamus driving appetite stimulation [1]. Ghrelin is another hormone produced in the

J.-P. H. Dedam et al., *Diseases of the Endocannabinoid System*,
https://doi.org/10.1007/978-3-031-87484-0_4

periphery, the lining of the stomach specifically, that also affects appetite in the hypothalamus. Its effect is to increase hunger as opposed to leptin, and this function is dependent on a functioning ECS. Blockage of the ECS in the hypothalamus prevents ghrelin from stimulating appetite [2]. The ECS is also involved in ghrelin production, which will be discussed below. The appetite-stimulating effect of corticosteroids in the hypothalamus is also made possible by increased production of endocannabinoids [3]. Yet another signal converging in the hypothalamus is the neuropeptide orexin. Orexin is best known for its role in increasing wakefulness and a mutation in its receptor leads to narcolepsy, but it also increases appetite. Endocannabinoids in the hypothalamus are able to greatly potentiate this function [4]. The ECS functions at many differ levels with the hypothalamus, boosting or blocking various inputs as well as facilitating others to overall drive greater food intake. Another site of action for driving food intake in the brain is the mesolimbic system, which controls emotional responses and reward. As was seen in the GI chapter, CB1 receptors in the nucleus accumbens strongly motivate food-seeking behaviors when activated, with increasing levels of 2-AG during fasting states [5].

It is important to point out, though, that chronic cannabis use is not actually associated with higher rates of obesity. The opposite is true, in fact, with reduced BMIs seen in chronic users, despite higher calorie intake [6]. This has been demonstrated in numerous studies including the large National Epidemiological Survey of Alcohol and Related Conditions in 2001–2002 [7]. Even though consuming cannabis does not cause obesity, the possibility that at least some cases of obesity are actually due to the body's own ECS dysfunction exists and will be explored in this chapter.

4.2 Gastric and Small Bowel Metabolic Signals

The ECS is also very active in the periphery to influence energy balance. CB1 receptors in the mucosal lining of the stomach provide another mechanism to increase food intake by stimulating the production of ghrelin from the enteroendocrine cells of the stomach. Ghrelin as discussed above is a hormone whose main role is increasing appetite. Patient who are undergoing gastric bypass have significantly lower ghrelin levels postoperatively, and this is thought to be one of the secondary effects of the surgery leading to weight loss [8]. Cholecystokinin (CCK) is another peptide, produced by enteroendocrine cells, that also influences food intake, more specifically this time in the small intestine. CCK, however, is released after eating and circulates back to the brain to act as a satiation signal. CB1 activation blocks the release of CCK and promotes continued feeding, despite adequate nutrient intake [9].

Once dietary fats are consumed, free fatty acids are sensed in the mouth and in turn activate the ECS in the small intestine, increasing endocannabinoid production via the vagal nerve. In one study in mice, feeding omega-6 fatty acids but not omega-3 and saturated fats caused increases in endocannabinoid levels [10]. Here, we have an instance of the endocannabinoid system departing from its typical role of being a feedback inhibitor and potentially providing a mechanism for driving an endocannabinoid excess syndrome. Omega-6 fatty acid intake in western diet is significantly higher than in previous human diets. This creates a feed-forward loop as more omega-6 fatty acids cause more endocannabinoid signaling, prompting greater appetite for more fatty foods containing omega-6 fatty acids. It is also possible that an increased production of endocannabinoids is further facilitated as they are made from omega-6 substrates not omega-3 fatty acids. Human studies show increased rates in obesity with increased omega-6 fatty acid diets [11]. This process may be driven in part by a feed-forward loop involving the ECS and omega-6 fatty acids, as described above.

4.3 Olfactory and Gustatory Controls

In mice, before food even enters their mouths, the ECS is active. In hungry mice, the ECS tone in the olfactory bulb is increased, leading to increased odor detection of food via CB1 signaling [12]. A small double-blinded study in humans, however, seemed to show the opposite; i.e., after 20 mg of THC was given to healthy volunteers, this resulted in decreased smell thresholds for detection and smell discrimination [13]. The question of what is happening to ECS levels in humans will likely go unanswered for some time as few volunteers are likely to line up to have their olfactory bulb biopsied. Taste is also affected by the ECS. AEA and 2-AG are known to increase response of taste receptor cells to sweet tastes but do not change response to salty, bitter, sour, or umami flavors. Leptin, once again, as an opposing physiologic signal, decreases reaction to sweet tastes [14].

4.4 Adipocyte Control

Blockage of CB1 receptor leading to weight loss goes beyond just reduction in food intake noted above. In mice models, leptin not only decreases endocannabinoid levels in the brain but also in adipocytes. Activation of CB1 receptors in adipocytes leads to increased fat production as well as decreased adiponectin, a hormone that improves insulin sensitivity and lowers inflammation. This was reversed by CB1 receptor antagonists [15]. Decreasing ECS tone in adipose cells has a direct effect, thereby decreasing fat storage. Cannabinoid receptors are present in white adipose tissues, which are involved in fat storage, and in brown adipose tissues, which burn fat to create heat. Although it has not been conclusively proven, there is some indication that the ECS is involved in the differentiation of white adipose cells from pro-adipose cells. CB2 receptors are not known to be present in adipocytes, and the small amount detected from fat samples may simply be related to infiltration of the tissues by immune cells [16]. Despite this, people with the CB2 receptor variant Q63R causing decreased receptor function have increased rates of obesity, which has also been replicated in mouse models. This would indicate an opposing function of CB2 receptor in the generation of obesity to CB1 receptor. Another single nucleotide change in CB1 receptor resulted in decreased BMI compared to controls in a large meta-analysis [17]. A different polymorphism in the CB1 receptor Rs806368, which appears to increase functions, resulted in increased central adiposity in a Brazilian population [18]. In an interested study, mesenchymal stem cells were taken from people with reduced CB2 function due to the above variant, and those stem cells were induced in vitro to become adipose cells and then studied. Adipose tissues are capable of producing a wide variety of adipokines, or cytokines (cell signaling proteins), produced by adipose tissues. We have already introduced two so far: leptin whose main function is to decrease food intake and adiponectin that reduces hepatic gluconeogenesis and improves insulin resistance. Patients with reduced CB2 receptor function had increased leptin and decreased adiponectin levels, the opposite of what occurs with reduced CB1 receptor function. Adipokines can also be inflammatory mediators like IL-6 and TNF that are partially responsible for the low levels of inflammation seen in obesity but also in people with reduced CB2 receptor function. The addition of a CB2 receptor agonist to boost CB2 activity reversed the above changes. People with low CB2 function also had decreased expression of UCP-1 (uncoupling protein), which is present in brown fat and allow the breakdown of fat to release heat [19]. White fat can be converted to brown fat by adding more mitochondria to break down that fat and convert it into heat. This process is blocked by CB1 receptor activation [20]. These findings indicate the important roles of both the ECS and immune system in obesity that require further study. CB2 stimulation, given its lack of psychoactive side effects, represents a much more attractive pharmacologic target than blockage of the CB1 receptor. Given the findings above, the overall increase in the ECS function could result in increase in obesity, as could an isolated decrease in CB2 function. This raises the possibility that increased adiposity could result from ECS dysfunction.

4.5 Hepatic Control and Metabolic Syndrome

In the liver, high-fat diets lead to insulin resistance, leptin resistance, and hyperglycemia. This unhealthy triad makes up the majority of what is commonly known as metabolic syndrome. The diagnosis of metabolic syndrome is made after at least three of the following changes are noted: abdominal obesity, high blood pressure, high blood sugar, high serum triglycerides, and low serum HDL (high-density lipoprotein). This process appears to be driven significantly by the ECS as mice without CB1 receptors do not develop this. Activation of CB1 receptor on liver cells leads to decreased insulin clearance by downregulating its degradation enzyme, which causes a hyper-insulin state and then insulin resistance [21]. Aside from high-fat diets, chronic stress is also thought to cause similar metabolic changes via high levels of circulating corticosteroids. Mice exposed to high levels of corticosteroids did not develop the expected metabolic changes when they either lacked CB1 receptors or had those receptors blocked. This study also showed that glucocorticoids increase endocannabinoid production in the liver as well as circulating levels in the blood [22]. And, as noted in the GI chapter, CB1 receptor activation on hepatocytes promotes lipogenesis as well, leading to increased fatty infiltration of the liver [23].

The clinical relevance of the ECS's effects on circulating lipids, however, is not well as well understood. Numerous studies have not shown significant changes in circulating triglyceride, HDL, or LDL levels in chronic cannabis users. However, many human studies looking at the effect of rimonabant, a CB1 receptor blocker, showed improvement in triglyceride levels and HDL [24]. One possible explanation for the discrepancy is that all of the human studies of rimonabant were done in patients with obesity and therefore likely some degree of ECS overactivation, thus benefiting them. Chronic cannabis use, on the other hand, does not seem to provoke a state of ECS overactivation and does not result in metabolic syndrome.

4.6 Pancreatic Control, Insulin Resistance and Type 2 Diabetes

In the pancreas, CB1 and CB2 receptors are present in islet cells, which produce insulin. Activation of these receptors leads to insulin hypersecretion, which could be a contributor to high levels of insulin seen in type 2 diabetes [25]. Activation of CB1 receptor in islet cells also inhibits insulin receptor activation, an autocrine effect whereby insulin stimulates the proliferation of islet cells. Blocking CB1 receptor in the pancreas results in greater islet cell mass and health [26]. The endocrine pancreatic function is also controlled in part by hormones called incretins. One of these glucagon-like peptide-1 is mimicked by the drug Ozempic, leading to increased insulin release but also preventing beta cell exhaustion, slowing GI transit, and decreasing appetite centrally. GLP-1 along with glucose -dependent insulinotropic polypeptide (GIP) is released from enteroendocrine cells in response to eating. These cells also express cannabinoid receptors. In a study in normal and obese subjects, GLP-1 and GIP levels were measured before and after the glucose challenge. The obese subjects had significantly high levels of GIP before and after the glucose challenge. Both groups had similar fasting GLP-1 levels, but the obese subjects had much lower GLP-1 levels after the glucose challenge. In the next phase of the study, health subjects were given a synthetic cannabinoid and then received the same glucose challenge. Activation of the ESC resulted in the same changes seen in obese subjects in the first part of the study: increased GIP throughout and decreased GLP-1 after the glucose challenge, indicating that changes in incretin secretion are likely driven by the ECS with unfavorable metabolic consequences [27]. A more recent study showed that for patients receiving a GLP-1 mimetic like Ozempic for the treatment of obesity, high pretreatment levels of endocannabinoids as OEA and PEA predicted greater response to the drug [28].

4.7 Skeletal Muscle Control

The vast majority of the energy consumption by our bodies occurs in skeletal muscles. Our muscle burn fat and glucose to produce the energy that moves our bodies. The degree of energy expenditure can be regulated by the ECS. On a cellular level,

AMPK acts as a "master switch" for this expenditure. The more AMPK present, the more fat is broken down to energy. CB1 activation in muscle cells decreases AMPK and therefore fat breakdown. Thus, the ECS can increase fatty stores by increasing the intake of fat in the diet but also decreasing its use in the largest consumer in the body [29]. In a study in mice using a selective CB1 receptor antagonist, mice treated with the blocker had decreased appetite and food intake as expected, but also showed increased oxygen consumption, indicating increased metabolism and also increased glucose intake by muscles cells [30]. Once again, the skeletal muscle is one of the biggest consumers of glucose in the body, and decreased glucose uptake in muscles due to insulin resistance is a major driver of hyperglycemia in diabetes. Here, again, we can see that overactivation of the ECS drives not only obesity but also potentially the development of type 2 diabetes and other metabolic diseases.

4.8 Anorexia Nervosa

Polymorphisms in genes encoding the FAAH enzyme, which break downs endocannabinoids, cause reduced function of the enzyme and increased circulating endocannabinoids as well as obesity [31]. Anandamide levels in both normal weight and obese human subjects peak while fasting. These levels then drop in normal weight subjects but not in obese subjects. No changes were noted in 2-AG in this study [32]. Eating food for pleasure, rather than dietary need, did cause increase in peripheral levels of ghrelin and 2-AG [33]. Underweight patients with anorexia nervosa, an eating disorder characterized by extreme weight loss, did not show increases in 2-AG in response to eating for pleasure [34]. A similar study in patients with anorexia nervosa found they had lower levels of anandamide compared to controls in fasting and postprandial states [35]. Both of these results along with the recent genetic study of polymorphisms in CB1 and CB2 genes causing increased risk of anorexia nervosa as well as lower BMIs in affected patients indicate the strong possibility of another endocannabinoid deficiency syndrome [36]. Several small studies using cannabinoids to treat anorexia nervosa have also been undertaken. One study, using 24 women with anorexia nervosa, administered 5 mg of dronabinol daily to half the women and half placebo for 4 weeks and then switched and gave dronabinol to the women who had received placebo for an additional 4 weeks. They were able to show a small improvement in weight in women getting dronabinol versus placebo [37].

4.9 Hemopressin, PEA, and OEA

Targeting the ECS as a means of treating obesity is still ongoing despite the failure of rimonabant. Peripherally restricted CB1 receptor antagonists have shown promise in mice but have not been pursued in human studies [38]. The discovery of the first endogenous peptide ligand of CB1 receptor, hemopressin, also presents additional therapeutic targets. Hemopressin is made by the proteolytic degradation of alpha hemoglobin. It acts as an inverse agonist of CB1 receptor, meaning that after binding of the cannabinoid receptor, it produces the opposite effect as would normally seen by binding of an endocannabinoid. As would be expected, it decreased feeding behaviors by acting in the hypothalamus without affecting reward pathways [39]. Hemopressin also appears to have other potentially beneficial effects on blood pressure, nociception, and inflammation, which are to be discussed in subsequent chapters.

Thus far in this chapter, we have focused on the effects of the ECS via the cannabinoid receptors. There are other receptors that are capable of being activated by endocannabinoid and endocannabinoid-related ligands like PEA and OEA as well as CBD. One in particular is very much involved in metabolism: PPAR gamma. PPAR (peroxisome proliferator-activated receptor) gamma is a nucleus receptor capable of regulating gene expression and is found extensively in adipose tissues, immune cells, and the colon. The effects of PPAR gamma activation are typically opposed to CB1 activation. Synthetic PPAR gamma ligands, from the drug class glitazones, have been used for years due to their insulin-sensitizing effects. In fat cells, they increase adiponectin and fatty acid breakdown. It is neuroprotective and also has an anti-inflammatory effect by decreasing the production of inflammatory cytokines and may be their underlaying mechanism to CBD's anti-inflammatory effects [40]. OEA can also exert an anorexic effect via PPAR alpha with beneficial lipid metabolism effects as well [41].

The role of the ECS in metabolism is immense, unfortunately in combination with the western diet, largely detrimental, and exacerbating if not outright, thereby driving obesity, metabolic syndrome, and type 2 diabetes. While chronic cannabis use does not further this process, person-to-person variations in the ECS may explain why some people become overweight and others do not, and the ongoing targeting of the endocannabinoid system for the treatment of metabolic disorders will only increase.

References

1. Di Marzo V, Goparaju SK, Wang L, Liu J, Bátkai S, Járai Z, Fezza F, Miura GI, Palmiter RD, Sugiura T, Kunos G. Leptin-regulated endocannabinoids are involved in maintaining food intake. Nature. 2001;410(6830):822–5. https://doi.org/10.1038/35071088.
2. Tucci SA, Rogers EK, Korbonits M, Kirlham TC. The cannabinoid receptor antagonist SR141716 blocks the orexigenic effect of intrahypothalamic ghrelin. Br J Pharmacol. 2004;143:520–3.
3. Malcher-Lopes R, Di S, Marcheselli VS, Weng FJ, Stuart CT, Bazan NG, Tasker JG. Opposing crosstalk between leptin and glucocorticoids rapidly modulates synaptic excitation via endocannabinoid release. J Neurosci. 2006;26:6643–50.
4. Hilairet S, et al. Hypersensitization of the orexin 1 receptor by the CB1 receptor: evidence for cross-talk blocked by the specific CB1 antagonist, SR141716. J Biol Chem. 2003;278(26):23731–7.
5. Kirkham TC, Williams CM, Fezza F, Di Marzo V. Endocannabinoid levels in rat limbic forebrain and hypothalamus in relation to fasting, feeding and satiation: stimulation of eating by 2-arachidonoyl glycerol. Br J Pharmacol. 2002;136:550–7.
6. Clark TM, Jones JM, Hall AG, Tabner SA, Kmiec RL. Theoretical explanation for reduced body mass index and obesity rates in *cannabis* users. Cannabis Cannabinoid Res. 2018;3(1):259–71. https://doi.org/10.1089/can.2018.0045.

7. Le Strat Y, Le Foll B. Obesity and cannabis use: results from 2 representative national surveys. Am J Epidemiol. 2011;174(8):929–33.
8. Senin LL, Al-Massadi O, Folgueira C, Castelao C, Pardo M, Barja-Fernandez S, Roca-Rivada A, Amil M, Crujeiras AB, Garcia-Caballero T, Gabellieri E, Leis R, Dieguez C, Pagotto U, Casanueva FF, Seoane LM. The gastric CB1 receptor modulates ghrelin production through the mTOR pathway to regulate food intake. PLoS One. 2013;8(11):e80339. https://doi.org/10.1371/journal.pone.0080339.
9. Argueta DA, Perez PA, Makriyannis A, DiPatrizio NV. Cannabinoid CB1 receptors inhibit gut-brain satiation signaling in diet-induced obesity. Front Physiol. 2019;11(10):704. https://doi.org/10.3389/fphys.2019.00704.
10. DiPatrizio NV, Joslin A, Jung KM, Piomelli D. Endocannabinoid signaling in the gut mediates preference for dietary unsaturated fats. FASEB J. 2013;27(6):2513–20. https://doi.org/10.1096/fj.13-227587.
11. Dayton S, Hashimoto S, Dixon W, Pearce ML. Composition of lipids in human serum and adipose tissue during prolonged feeding of a diet high in unsaturated fat. J Lipid Res. 1966;7(1):103–11.
12. Soria-Gómez E, Bellocchio L, Reguero L, Lepousez G, Martin C, Bendahmane M, Ruehle S, Remmers F, Desprez T, Matias I, Wiesner T, Cannich A, Nissant A, Wadleigh A, Pape HC, Chiarlone AP, Quarta C, Verrier D, Vincent P, Massa F, Lutz B, Guzmán M, Gurden H, Ferreira G, Lledo PM, Grandes P, Marsicano G. The endocannabinoid system controls food intake via olfactory processes. Nat Neurosci. 2014 Mar;17(3):407–15. https://doi.org/10.1038/nn.3647. Epub 2014 Feb 9. PMID: 24509429.
13. Walter C, Oertel BG, Ludyga D, Ultsch A, Hummel T, Lötsch J. Effects of 20 mg oral Δ(9)-tetrahydrocannabinol on the olfactory function of healthy volunteers. Br J Clin Pharmacol. 2014 Nov;78(5):961–9. https://doi.org/10.1111/bcp.12415. PMID: 24802974; PMCID: PMC4243870.
14. Yoshida R, Niki M, Jyotaki M, Sanematsu K, Shigemura N, Ninomiya Y. Modulation of sweet responses of taste receptor cells. Semin Cell Dev Biol. 2013;24(3):226–31. https://doi.org/10.1016/j.semcdb.2012.08.004.
15. Matias I, Gonthier M-P, Orlando P, Martiadis V, De Petrocellis L, Cervino C, Petrosino S, Hoareau L, Festy F, Pasquali R, Roche R, Maj M, Pagotto U, Monteleone P, Di Marzo V. Regulation, function, and dysregulation of endocannabinoids in models of adipose and β-pancreatic cells and in obesity and hyperglycemia. J Clin Endocrinol Metabol. 2006;91(8):3171–80.
16. Jung KM, Lin L, Piomelli D. The endocannabinoid system in the adipose organ. Rev Endocr Metab Disord. 2022;23(1):51–60. https://doi.org/10.1007/s11154-020-09623-z.
17. Sadeghian M, Rahmani S, Mansoori A. G1359A variant of the cannabinoid receptor gene (rs1049353) and obesity-related traits and related endophenotypes: a meta-analysis. Ann Nutr Metab. 2018;73(1):76–85. https://doi.org/10.1159/000490668.
18. Jaeger JP, Mattevi VS, Callegari-Jacques SM, Hutz MH. Cannabinoid type-1 receptor gene polymorphisms are associated with central obesity in a Southern Brazilian population. Dis Markers. 2008;25(1):67–74. https://doi.org/10.1155/2008/841490.
19. Rossi F, Bellini G, Luongo L, Manzo I, Tolone S, Tortora C, et al. Cannabinoid receptor 2 as Antiobesity target: inflammation, fat storage, and Browning modulation. J Clin Endocrinol Metab. 2016;101(9):3469–78.
20. Tedesco L, Valerio A, Dossena M, Cardile A, Ragni M, Pagano C, Pagotto U, Carruba MO, Vettor R, Nisoli E. Cannabinoid receptor stimulation impairs mitochondrial biogenesis in mouse white adipose tissue, muscle, and liver: the role of eNOS, p38 MAPK, and AMPK pathways. Diabetes. 2010;59(11):2826–36. https://doi.org/10.2337/db09-1881.
21. Liu J, et al. Hepatic cannabinoid receptor-1 mediates diet-induced insulin resistance via inhibition of insulin signaling and clearance in mice. Gastroenterology. 2012;142(5):1218–28.
22. Bowles NP, Karatsoreos IN, Li X, Vemuri VK, Wood JA, Li Z, Tamashiro KL, Schwartz GJ, Makriyannis AM, Kunos G, Hillard CJ, McEwen BS, Hill MN. A peripheral endocannabinoid mechanism contributes to glucocorticoidmediated metabolic syndrome. Proc Natl Acad Sci

U S A. 2015 Jan 6;112(1):285–90. https://doi.org/10.1073/pnas.1421420112. Epub 2014 Dec 22. PMID: 25535367; PMCID: PMC4291642.

23. Maccarrone M, Bab I, Bíró T, Cabral GA, Dey SK, Di Marzo V, Konje JC, Kunos G, Mechoulam R, Pacher P, Sharkey KA, Zimmer A. Endocannabinoid signaling at the periphery: 50 years after THC. Trends Pharmacol Sci. 2015;36(5):277–96. https://doi.org/10.1016/j.tips.2015.02.008.
24. Lazarte J, Hegele RA. Cannabis effects on lipoproteins. Curr Opin Lipidol. 2019;30(2):140–6.
25. Malenczyk K, et al. CB1 cannabinoid receptors couple to focal adhesion kinase to control insulin release. J Biol Chem. 2013;288(45):32685–99.
26. Kim W, et al. Cannabinoids inhibit insulin receptor signaling in pancreatic β-cells. Diabetes. 2011;60(4):1198–209.
27. Chia CW, Carlson OD, Liu DD, González-Mariscal I, Santa-Cruz Calvo S, Egan JM. Incretin secretion in humans is under the influence of cannabinoid receptors. Am J Physiol Endocrinol Metab. 2017;313(3):E359–66. https://doi.org/10.1152/ajpendo.00080.2017.
28. Matias I, Lehmann EW, Zizzari P, Byberg S, Cota D, Torekov SS, Quarta C. Endocannabinoid-related molecules predict the metabolic efficacy of GLP-1 receptor agonism in humans with obesity. J Endocrinol Investig. 2023;47:1289. https://doi.org/10.1007/s40618-023-02228-8.
29. Cavuoto P, et al. Effects of cannabinoid receptors on skeletal muscle oxidative pathways. Mol Cell Endocrinol. 2007;267(1–2):63–9.
30. Liu YL, et al. Effects of the cannabinoid CB1 receptor antagonist SR141716 on oxygen consumption and soleus muscle glucose uptake in Lepob/Lepob mice. Int J Obes. 2005;29(2):183–7.
31. Sipe JC, et al. Overweight and obesity associated with a missense polymorphism in fatty acid amide hydrolase (FAAH). Int J Obes. 2005;29(7):755–9.
32. Gatta-Cherifi B, et al. Simultaneous postprandial deregulation of the orexigenic endocannabinoid anandamide and the anorexigenic peptide YY in obesity. Int J Obes. 2012;36(6):880–5.
33. Monteleone P, et al. Hedonic eating is associated with increased peripheral levels of ghrelin and the endocannabinoid 2-arachidonoyl-glycerol in healthy humans: a pilot study. J Clin Endocrinol Metabol. 2012;97(6):E917–24.
34. Monteleone AM, et al. Deranged endocannabinoid responses to hedonic eating in underweight and recently weight-restored patients with anorexia nervosa. Am J Clin Nutr. 2015;101(2):262–9.
35. Piccolo M, Claussen MC, Bluemel S, Schumacher S, Cronin A, Fried M, Goetze O, Martin-Soelch C, Milos G. Altered circulating endocannabinoids in anorexia nervosa during acute and weight-restored phases: a pilot study. Eur Eat Disord Rev. 2020;28(1):46–54. https://doi.org/10.1002/erv.2709.
36. González LM, García-Herráiz A, Mota-Zamorano S, Flores I, Albuquerque D, Gervasini G. Variability in cannabinoid receptor genes is associated with psychiatric comorbidities in anorexia nervosa. Eat Weight Disord. 2021;26(8):2597–2606. https://doi.org/10.1007/s40519-021-01106-7. Epub 2021 Feb 11. PMID: 33575982.
37. Andries A, Frystyk J, Flyvbjerg A, Støving RK. Dronabinol in severe, enduring anorexia nervosa: a randomized controlled trial. Int J Eat Disord. 2014;47(1):18–23. https://doi.org/10.1002/eat.22173.
38. Cluny NL, et al. A novel peripherally restricted cannabinoid receptor antagonist, AM6545, reduces food intake and body weight, but does not cause malaise, in rodents. Br J Pharmacol. 2010;161(3):629–42.
39. Dodd GT, Worth AA, Hodkinson DJ, Srivastava RK, Lutz B, Williams SR, Luckman SM. Central functional response to the novel peptide cannabinoid, hemopressin. Neuropharmacology. 2013;71:27–36. https://doi.org/10.1016/j.neuropharm.2013.03.007.
40. Marion-Letellier R, Savoye G, Ghosh S. Fatty acids, eicosanoids and PPAR gamma. Eur J Pharmacol. 2016;15(785):44–9. https://doi.org/10.1016/j.ejphar.2015.11.004.
41. Fu J, Oveisi F, Gaetani S, Lin E, Piomelli D. Oleoylethanolamide, an endogenous PPAR-alpha agonist, lowers body weight and hyperlipidemia in obese rats. Neuropharmacology. 2005;48(8):1147–53. https://doi.org/10.1016/j.neuropharm.2005.02.013.

5 Cardiovascular System

The effects of the ECS on our cardiovascular system are well studied but in many senses are still not well understood. Large population studies in Canada following patients with medical cannabis authorization linked cannabis use with greater risk of heart attack, but a separate large study using the UK Biobank, arguably a similar patient population, has shown a reduction in MI similar to that of red wine consumption [1]. As in all cases with the ECS, the truth is complicated. CB1 receptors are found extensively in the heart and vascular tissues. They are present in the central regulatory centers of the CV system in the brain, as well in the sympathetic and parasympathetic nerves that link the heart and blood vessels to central regulatory centers [2].

5.1 Effect on Myocardial Cells

Anandamide was shown in experiments on rat myocardial cells to decrease action potential duration and amplitude via L-type calcium channels and CB1 receptor activation, potentially having an antiarrhythmic effect in addition to the decrease in contractility [3, 4]. Subsequent studies have shown anandamide's ability to inhibit L-type calcium channels and voltage-dependent sodium channels via a mechanism independent of either cannabinoid receptor [5]. The negative inotropic effects of CB1 receptor have also been confirmed in experiments using human atrial muscles [6]. CB1 receptors are also present in the sinoatrial node pacemaker cells in rabbits where they also decreased action potential duration and amplitude via calcium channel but not potassium modulation [7]. This may explain observational studies using the National Inpatient Survey database, suggesting that cannabis users were less likely to have atrial fibrillation, although the effect was small: 19% versus 21% [8]. Activation of the same receptor on vascular smooth muscles leads to vasodilation and also decreases the influx of calcium via the same L-type calcium channel [9].

J.-P. H. Dedam et al., *Diseases of the Endocannabinoid System*,
https://doi.org/10.1007/978-3-031-87484-0_5

5.2 Central Control of CV System

Centrally, the ECS appears to interact significantly with the brain renin-angiotensin system modulating the autonomic nerve output at a number of different brain nuclei. Briefly, the enzyme renin is produced in the kidneys in response to low blood pressure and released into the blood stream. Renin then converts angiotensinogen from the liver to angiotensin I, which is converted to angiotensin II, mainly in the lungs. Angiotensin II has many effects throughout the body including stimulation of adrenal glands to produce aldosterone and vasoconstriction of blood vessels and in the brain including increasing the sympathetic output and the production of antidiuretic hormone in the pituitary gland. Activation of CB1 receptor in the paraventricular nucleus (PVN) of the thalamus, an important control center for blood pressure, can have a bidirectional effect on blood pressure. Activation of the ECS here is necessary to mediate the hypertensive effects of angiotensin II [10]. Blockage of CB1 receptors in mice with angiotensin-driven hypertensive rats caused significant drops in blood pressure [11]. However, direct injection of CB1 receptor agonists into this region of the brain in mice results in lower blood pressure and heart rate. This seemingly contradictory effect is due to the inhibitory effect of CB1 receptors presynaptically to both glutamatergic and GABA neurons present in this region of the brain. Glutamatergic neurons here drive increased sympathetic nerve output and GABA neurons reduce it. Inhibition of the glutamatergic neurons decreases the sympathetic output and causes a drop in blood pressure and heart rate. Inhibition of GABA neurons allows increased sympathetic tone and results in increased blood pressure and heart rate. The balance of which neurons are inhibited or not and whether blood pressure and heart rates go up or down appears to be due to the local presence of a variety of signals including angiotensin II, catecholamines, and nitric oxide [12].

The excitatory glutamatergic neurons of the PVN then project down to the rostral ventrolateral medulla (RVLM) where they stimulate, at least in part, the sympathetic nerve output [13]. CB1 receptor activation in the rostral ventrolateral medulla in rats also increases renal sympathetic nerve activity and causes an increase in blood pressure [14, 15]. The same region of the brain in humans is similarly involved in the basal and reflex control sympathetic control of the cardiovascular system, specifically increasing heart rate and blood pressure in response to position change. The recently characterized endocannabinoid receptor GPR18 is also expressed here and in rat studies had a tonic blood pressure-lowering effect. This BP-lowering effect of GPR18, however, was mitigated by the activation of CB1 receptor in the same region [16]. It would appear then that dominant effect of the ECS in this region of the brain is to lower blood pressure, except when needed for reflexive increases due to sudden position changes like sitting to standing.

Also present in the medulla is the solitary tract nucleus (NTS). Here, the role of the ECS is even more confusing. Within the NTS, parasympathetic fibers of the vagus nerve and sympathetic nerve outputs are involved in cardiac baroreflex sensitivity. Increases in blood pressure lead to increased parasympathetic tone, decreasing heart rate and inhibiting sympathetic tone to lower blood pressure. Increases in blood pressure result in elevated anandamide levels in the NTS. Injection of AEA into this region of the brain results in decreased sympathetic output, and blockage of CB1 receptors increases the parasympathetic output [17]. Here, therefore, ECS appears to inhibit all autonomic activities. This effect on decreasing the cardiac baroreflex may explain some of THC's cardiac side effects as well as reported orthostatic hypotension after use. THC is known to increase heart rate possibly though suppression of parasympathetic tone as described above while at the same time decreasing blood pressure, via vasodilation but also potentially through decreased sympathetic tone from the NTS, which is also described above.

The pepcan hemopressin is also known to result in a slight drop in blood pressure in rodents, an effect that may be mediated via its interaction with the ECS [18].

5.3 Parasympathetic and Sympathetic Nerve-Mediated Control

CB1 receptors also seem to have an effect directly on parasympathetic and sympathetic nerve inputs on the heart itself. In experiments using rabbits, activation of CB1 inhibited cardio-acceleration. This occurred by inhibition of norepinephrine release in the superior cervical ganglion but also presynaptically in the heart itself. Interestingly, CB1 receptor activation also inhibited vagal activity in the heart, resulting in tachycardia. It is unclear if this is occurring in the dorsal motor nucleus of the vagus, which expresses CB1, or presynaptically on the vagal inputs to the heart itself [19]. THC generally results in tachycardia in naive subjects, an effect that can be blocked by atropine and propranolol, indicating the involvement of both sympathetic and parasympathetic inputs to create the effect [20]. The same effect was seen in experienced users with increases in heart rate correlating with increased dose [21]. Once again, we see THC increasing heart rate through blockade of parasympathetic activity in the heart while at the same time decreasing sympathetic tone to prevent an increase in blood pressure. This may also play a role in finding decreased rates of atrial fibrillation in patient who use cannabis mentioned previously in the chapter, as increased sympathetic tone increases the risk of atrial fibrillation.

5.4 ECS and Hypertension

CB1 knockout mice do not have any significant observable baseline changes in blood pressure or heart rate. However, there is some evidence for a role of the ECS in hypertensive disorders.

Anandamide levels are increased in patients with essential hypertension as well as hypertensive patients with obstructive sleep apnea but not in women with pre-eclampsia, a hypertensive disorder of pregnancy [3, 4, 22, 23]. Anandamide when given to anesthetized rats has a triphasic cardiovascular effect: first, a transient hypotension due to increased vagal activity; second, a brief increase in blood pressure; and third, a prolonged period of lower blood pressure. The last two stages are also observed with THC administration. Blockage of CB1 as well as alpha-adrenergic blockade prevented the latter low blood pressure period but not the brief increase in blood pressure [24]. Further studies have shown that blockage of beta-2 adrenergic centrally can prevent the anandamide pressor effect, and blockage of NMDA and thromboxane A (2) receptors peripherally also prevented the pressor effect [25]. Analysis of data from the US National Health and Nutrition Examination Surveys (2005–2012) with over 12,000 participants found an increase in systolic blood pressure of 1.6 mmHg in patients who admitted to cannabis use within the last 30 days [26]. Other studies looking at cardiac risk factors have not shown a significant association been cannabis use and elevated blood pressure. Perhaps admitting to the government that you were a cannabis user was sufficient to increase blood pressure in respondents. A large study in England of more than 90,000 participants found current cannabis users had slightly lower systolic blood pressures: 0.63 mmHg less in men and 1.17 less in women [27]. Numerous randomized controlled trials looking at various cannabinoids used in medical care such as nabilone, a synthetic cannabinoid partial agonist of CB1 and CB2 receptors, reported low blood pressure as a side effect, showing a pooled relative risk of 3.47% for orthostatic hypotension and 3.14% for hypotension in general [28].

Peripherally, there is also another interaction between angiotensin II and the ECS. Angiotensin II causes vasoconstriction in mouse and rat blood vessels while at the same time prompting the production of 2-AG. 2-AG via CB1 receptor then attenuated the effect of angiotensin II by causing vasodilation [29]. The increase in serum anandamide noted in several hypertensive disorders mentioned above may be the result of a feedback inhibition mechanism as described above with the interaction between angiotensin II and 2-AG or simply an overactivation of the ECS known to occur in metabolic syndrome and discussed in detail in the chapter of metabolism. The possibility that increased ECS tone in hypertension is an attempt by the body to feedback and ultimately lower blood pressure is bolstered by the observation that inhibition of FAAH, the enzyme that breaks down anandamide in the experiment above, normalized the blood pressure of hypertensive rats. In studies of rats, the blood pressure-lowering effect of CB1 receptor was much greater in rats with hypertension than in rats with normal blood pressures, mediated by both vasodilation and decreased cardiac contractility [30]. The site of the vasodilation was predominant in coronary and cerebral vessels but also in the mesentery and is accomplished through reduction of sympathetic tone [31, 32].

5.5 TRPV1 and PPAR

Two other receptors that are known to be activated by the ECS and play a role in the cardiovascular system are the transient receptor potential vanilloid channel subfamily member 1 or TRPV1 and peroxisome proliferator-activated receptor or PPAR, specifically alpha and gamma types. TRPV1 is expressed thorough the CV system: sensory nerve fibers in the myocardium, the myocardium itself, and endothelial and vascular smooth muscle cells of vessels and platelets. Activation leads to nitric oxide, calcitonin gene-related protein, and substance P production [2]. It has some beneficial effects during MI and atherosclerosis, which will be discussed further below, but this is tempered by its negative effects in driving metabolic syndrome, similar to the rest of the ECS, that ultimately resulted in negative CV outcomes in mice models. [33]. Activation of PPAR appears to reduce the overall blood pressure peripherally by reducing IL-6 and Cox-2 inflammatory markers via PPAR alpha as well as centrally via PPAR gamma and the renin-angiotensin brain axis [34, 35]. Taken all together, an overall increase in ECS could result in elevated BP; however, given the number of opposing/counterbalancing effects within ECS itself both centrally and peripherally, this effect would be small. And, as was mentioned before, CB1 knockout mice do not have a significant change in BP, indicating that ECS deficiency would likely not affect BP either.

5.6 ECS and MI

Activation of the ECS also occurs after myocardial infarction (MI). During MI, or any major injury to the ventricles, there can occur a cardioprotective reflex known as the von Bezold-Jarisch reflex, with increased parasympathetic tone resulting in bradycardia, vasodilation, and hypotension mediated by activation of TRPV1 receptors on vagus sensory nerves of the myocardium. Rat models of MI have found this mechanism to limit the size of MI [36]. Increased levels of anandamide have been found in coronary arteries following acute MI. Patients with detectable anandamide levels at the time of percutaneous coronary intervention, the gold standard treatment of occluded arteries in the heart, had better outcomes than those who did not [37]. In rat models of MI, increased levels of anandamide and 2-AG produced in circulating platelets and monocytes mediated hypotension and bradycardia, and this effect was reduced by a selective CB1 receptor blocker. Blockage of CB1 receptor in these rats also resulted in increased mortality [31, 32]. Other potential beneficial effects of the ECS after MI may also include advantageous metabolic changes in myocytes and prevention of apoptosis; however, both require further study. After an MI, inflammation can lead to ischemic reperfusion injury. Activation of CB2 receptor in mice models has been shown to decrease further tissue injury following MI [38]. Long-term activation of CB2 receptors by THC in mouse models also limited progression of atherosclerosis, the inflammatory process that causes the MIs in the first place [39]. As seen in the small human studies above and based on animal models, ECS deficiency would certainly result in worsened post-MI outcomes.

5.7 ECS and Atherosclerosis

MIs are the end result of atherosclerosis, the slower process of artery thickening, accumulation of oxidized LDL in the artery wall, and eventually narrowing or stenosis of that artery. These fatty plaques, made up of foam cells, can either slowly restrict blood flow to target organs and cause ischemia or rupture spontaneously completing blocking flow of blood resulting in MI in the heart or stroke in the brain. Here again, we see a somewhat contradictory role of the ECS. When overactivated systemically as discussed at length in the chapter of metabolism, the ECS drives metabolic syndrome, a major risk factor of atherosclerosis development. But studies in mice with the FAAH gene knocked out and higher levels of AEA showed they had less age-related cardiovascular disease than wild-type mice, a process driven largely by less chronic inflammation [40]. Human studies blocking CB1 receptor with the drug rimonabant did not improve progression of atherosclerotic disease in two separate studies, despite weight loss [41, 42]. One of the early steps in development of atherosclerosis is endothelial dysfunction and injury. This results in inflammation that attracts monocytes from the blood that attach and invade the walls of vessels, consuming oxidized LDL to become the enlarged foam cells of atherosclerotic plaques. In a study using the epithelial cells of human coronary arteries, CB1 receptor activation contributed to cell death via generation of reactive oxygen species [43]. Activation of CB2 receptor, on the other hand, decreased inflammation via decreased TNF-alpha production, leading to decreased migration of monocytes [44]. CB2 activation was also credited in a study above showing limited progression of atherosclerosis in mice treated with long-term THC. The mechanisms proposed was inhibition of migration and proliferation of T cells into vessel walls, another initiating step in the development of atherosclerosis [39]. TRPV1 also has beneficial effects here with studies demonstrating decreased inflammation in vascular epithelial cells via nitric oxide production and impeded foam cell development by the vascular smooth muscle cells of mice [45, 46]. PPAR alpha and gamma activation has similar effects to CB2 activation, which also decreases inflammation of macrophage foam cell formation [47]. Taken together, it seems very unlikely that the ECS plays a role in driving atherosclerosis, and its overall role in the body is prevention. A relative deficiency in CB2 receptor activity would seem likely to result in higher risk of heart disease; however, a large German population study looking at common CB2 polymorphisms did not demonstrate any increased risk [48]. Whether selective activation of CB2 receptors prior to or following MI would have benefit, however, remains to be tested.

5.8 Cannabis and MI

There are many studies now looking at the association between cannabis and heart disease, again with mixed results. In the CARDIA study, an observational study of 3498 participants, cannabis use was only associated with subclinical atherosclerosis, i.e., coronary artery calcium, in participants who were also using tobacco [49]. The study also did not find a cumulative lifetime risk or recent use of cannabis was associated with cardiovascular diseases [50]. Cannabis does appear to be causing increased admissions for acute MI and arrhythmia in younger people (18–39 years old) based on the review of the National Inpatient Sample. However, they had a slightly lower all-cause mortality compared to non-cannabis users who were admitted [51]. In this study showing increased admission for MI, they did not differentiate between type 1 MI caused by actually blockage of the arteries and type II MI caused by ischemic demand without the actual blockage of the artery. Type II MIs are much more common in the age group as very few at this age have actually developed coronary diseases yet. Type II MIs also do not have as high a morbidity or mortality as type I MIs. The type of arrhythmia is also not defined and could include sinus tachycardia or any heart rate over 90 beats per minute, a known side effect of cannabis. Cannabis's side effect of hypotension in combination with other physiologic stresses could also result in ischemic demand and increased occurrences of type II MIs. When using the same National Inpatient Survey to look at all ages and gender, patients with an ICD 9 diagnosis of recreational cannabis use had an odds ratio of 1.079 for MI diagnosis compared to nonusers [52]. This once again did not differentiate between type 1 and type 2 MIs, nor did they consider the potential bias of people of color or people diagnosed with mental illness being more likely to "diagnosed" with cannabis use given the associated stigma. In fact, a large retrospective study from eight US states looking at over 1 million hospital admissions found cannabis users were less likely to have coronary diseases and had lower inhospital mortality [53]. The large study in Canada mentioned at the start of the chapter showed an increase hazard ratio for patients with medical cannabis authorization for both acute coronary syndrome and stroke. The study did not adjust for concurrent tobacco use, however. None of the studies above looked at the effect of smokable versus non-smokable cannabis on MI and heart disease as for the act of smoking, anything will have negative CV outcomes.

5.9 ECS and Heart Failure

Another major cause of cardiac morbidity and mortality is heart failure. Studies in mice with CB1 receptor deficiency had significantly higher mortality rates in heart failure models, and this was attributed to excessive sympathetic activation [54]. These mice where then followed and were shown to experience worse cardiac remodeling than wild-type mice. They had higher left ventricular end-diastolic volumes, a marker for worse congestive heart failure as lower left ventricular contractility and larger hearts overall [55, 56]. These effects have borne out in human observation studies using the National Inpatient Sample database, showing that the use of cannabis among patients diagnosed with heart failure had decreased inpatient mortality and decreased length of hospital stay [57]. Here, as in MI, a relative ECS could very well result in worsened outcomes in heart failure.

While the role of the ECS in hypertension seems more mixed, it is vital for reducing the life-threatening complications of MI and heart failure. The association between cannabis use and heart disease is also still somewhat muddled. Perhaps, cannabis biggest risk to cardiovascular health is its well-documented association with tobacco smoking and alcohol and drug use that are known to clearly cause increased heart disease [58, 59].

References

1. Lehrer S, Rheinstein PH. Marijuana and myocardial infarction in the UK biobank cohort. Cureus. 2022;14(2):e22054. https://doi.org/10.7759/cureus.22054.
2. Sierra S, Luquin N, Navarro-Otano J. The endocannabinoid system in cardiovascular function: novel insights and clinical implications. Clin Auton Res. 2018;28(1):35–52. https://doi.org/10.1007/s10286-017-0488-5.
3. Li Q, Ma HJ, Zhang H, Qi Z, Guan Y, Zhang Y. Electrophysiological effects of anandamide on rat myocardium. Br J Pharmacol. 2009a;158(8):2022–9. https://doi.org/10.1111/j.1476-5381.2009.00518.x.
4. Li D, Chen BM, Peng J, Zhang YS, Li XH, Yuan Q, Hu CP, Deng HW, Li YJ. Role of anandamide transporter in regulating calcitonin gene-related peptide production and blood pressure in hypertension. J Hypertens. 2009b;27(6):1224–32. https://doi.org/10.1097/hjh.0b013e328329bbd7.
5. Al Kury LT, Voitychuk OI, Yang KH, Thayyullathil FT, Doroshenko P, Ramez AM, Shuba YM, Galadari S, Howarth FC, Oz M. Effects of the endogenous cannabinoid anandamide on voltage-dependent sodium and calcium channels in rat ventricular myocytes. Br J Pharmacol. 2014;171(14):3485–98. https://doi.org/10.1111/bph.12734.
6. Bonz A, Laser M, Küllmer S, Kniesch S, Babin-Ebell J, Popp V, Ertl G, Wagner JA. Cannabinoids acting on CB1 receptors decrease contractile performance in human atrial muscle. J Cardiovasc Pharmacol. 2003;41(4):657–64. https://doi.org/10.1097/00005344-200304000-00020.
7. Zhang J, Wang SY, Zhou JJ, Wei Y, Li Q, Yang J, Zhang Y. Inhibitory effects of endocannabinoid on the action potential of pacemaker cells in sinoatrial nodes of rabbits. Sheng Li Xue Bao. 2013;65(2):129–34.
8. Adegbala O, Adejumo AC, Olakanmi O, Akinjero A, Akintoye E, Alliu S, Edo-Osagie E, Chatterjee A. Relation of cannabis use and atrial fibrillation among patients hospitalized for heart failure. Am J Cardiol. 2018;122(1):129–34. https://doi.org/10.1016/j.amjcard.2018.03.015.
9. Gebremedhin D, Lange AR, Campbell WB, Hillard CJ, Harder DR. Cannabinoid CB1 receptor of cat cerebral arterial muscle functions to inhibit L-type Ca2+ channel current. Am J Phys. 1999;276(6):H2085–93. https://doi.org/10.1152/ajpheart.1999.276.6.H2085.
10. Gyombolai P, Pap D, Turu G, Catt KJ, Bagdy G, Hunyady L. Regulation of endocannabinoid release by G proteins: a paracrine mechanism of G protein-coupled receptor action. Mol Cell Endocrinol. 2012;353:29–36. https://doi.org/10.1016/j.mce.2011.10.011.
11. Schaich CL, Shaltout HA, Brosnihan KB, Howlett AC, Diz DI. Acute and chronic systemic CB1 cannabinoid receptor blockade improves blood pressure regulation and metabolic profile in hypertensive (mRen2)27 rats. Physiol Rep. 2014;2(8):e12108. https://doi.org/10.14814/phy2.12108.
12. Grzęda E, Schlicker E, Toczek M, Zalewska I, Baranowska-Kuczko M, Malinowska B. CB1 receptor activation in the rat paraventricular nucleus induces bi-directional cardiovascular effects via modification of glutamatergic and GABAergic neurotransmission. Naunyn Schmiedeberg's Arch Pharmacol. 2017;390(1):25–35. https://doi.org/10.1007/s00210-016-1302-y. Epub 2016 Sep 22
13. Koba S, Hanai E, Kumada N, Kataoka N, Nakamura K, Watanabe T. Sympathoexcitation by hypothalamic paraventricular nucleus neurons projecting to the rostral ventrolateral medulla. J Physiol. 2018;596(19):4581–95. https://doi.org/10.1113/JP276223.
14. Wang T, Li GQ, Zhang HP, Zhang Y, Li Q. Overactivation of cannabinoid receptor type 1 in rostral ventrolateral medulla promotes cardiovascular responses in spontaneously hypertensive rats. J Hypertens. 2017a Mar;35(3):538–45. https://doi.org/10.1097/HJH.0000000000001179.
15. Wang Y, Cui L, Xu H, Liu S, Zhu F, Yan F, Shen S, Zhu M. TRPV1 agonism inhibits endothelial cell inflammation via activation of eNOS/NO pathway. Atherosclerosis. 2017b;260:13–9. https://doi.org/10.1016/j.atherosclerosis.2017.03.016.

16. Penumarti A, Abdel-Rahman AA. The novel endocannabinoid receptor GPR18 is expressed in the rostral ventrolateral medulla and exerts tonic restraining influence on blood pressure. J Pharmacol Exp Ther. 2014;349(1):29–38. https://doi.org/10.1124/jpet.113.209213.
17. Schaich CL, Grabenauer M, Thomas BF, Shaltout HA, Gallagher PE, Howlett AC, Diz DI. Medullary endocannabinoids contribute to the differential resting Baroreflex sensitivity in rats with altered brain renin-angiotensin system expression. Front Physiol. 2016;9(7):207. https://doi.org/10.3389/fphys.2016.00207.
18. Blais PA, Côté J, Morin J, Larouche A, Gendron G, Fortier A, Regoli D, Neugebauer W, Gobeil F Jr. Hypotensive effects of hemopressin and bradykinin in rabbits, rats and mice. A comparative study. Peptides. 2005 Aug;26(8):1317–22. https://doi.org/10.1016/j.peptides.2005.03.026. Epub 2005 Apr 26. PMID: 16042973.
19. Szabo B, Nordheim U, Niederhoffer N. Effects of cannabinoids on sympathetic and parasympathetic neuroeffector transmission in the rabbit heart. J Pharmacol Exp Ther. 2001;297(2):819–26.
20. Benowitz NL, Rosenberg J, Rogers W, Bachman J, Jones RT. Cardiovascular effects of intravenous delta-9-tetrahydrocannabinol: autonomic nervous mechanisms. Clin Pharmacol Ther. 1979;25(4):440–6. https://doi.org/10.1002/cpt1979254440.
21. Perez-Reyes M, Di Guiseppi S, Davis KH, Schindler VH, Cook CE. Comparison of effects of marihuana cigarettes to three different potencies. Clin Pharmacol Ther. 1982;31(5):617–24. https://doi.org/10.1038/clpt.1982.86.
22. Engeli S, Blüher M, Jumpertz R, Wiesner T, Wirtz H, Bosse-Henck A, Stumvoll M, Batkai S, Pacher P, Harvey-White J, Kunos G, Jordan J. Circulating anandamide and blood pressure in patients with obstructive sleep apnea. J Hypertens. 2012;30(12):2345–51. https://doi.org/10.1097/HJH.0b013e3283591595.
23. Molvarec A, Fügedi G, Szabó E, Stenczer B, Walentin S, Rigó J Jr. Decreased circulating anandamide levels in preeclampsia. Hypertens Res. 2015;38(6):413–8. https://doi.org/10.1038/hr.2015.20.
24. Lake KD, Martin BR, Kunos G, Varga K. Cardiovascular effects of anandamide in anesthetized and conscious normotensive and hypertensive rats. Hypertension. 1997;29(5):1204–10. https://doi.org/10.1161/01.hyp.29.5.1204.
25. Malinowska B, Zakrzeska A, Kurz CM, Göthert M, Kwolek G, Wielgat P, Braszko JJ, Schlicker E. Involvement of central beta2-adrenergic, NMDA and thromboxane A2 receptors in the pressor effect of anandamide in rats. Naunyn Schmiedeberg's Arch Pharmacol. 2010;381(4):349–60. https://doi.org/10.1007/s00210-010-0497-6.
26. Alshaarawy O, Elbaz HA. Cannabis use and blood pressure levels: United States National Health and nutrition examination survey, 2005-2012. J Hypertens. 2016;34(8):1507–12. https://doi.org/10.1097/HJH.0000000000000990.
27. Vallée A. Association between cannabis use and blood pressure levels according to comorbidities and socioeconomic status. Sci Rep. 2023;13:2069. https://doi.org/10.1038/s41598-022-22841-6.
28. Watanabe AH, Navaravong L, Sirilak T, Prasitwarachot R, Nathisuwan S, Page RL 2nd, Chaiyakunapruk N. A systematic review and meta-analysis of randomized controlled trials of cardiovascular toxicity of medical cannabinoids. J Am Pharm Assoc. 2003;61(5):e1–e13. https://doi.org/10.1016/j.japh.2021.03.013.
29. Szekeres M, Nádasy GL, Turu G, Soltész-Katona E, Tóth ZE, Balla A, Catt KJ, Hunyady L. Angiotensin II induces vascular endocannabinoid release, which attenuates its vasoconstrictor effect via CB1 cannabinoid receptors. J Biol Chem. 2012;287(37):31540–50. https://doi.org/10.1074/jbc.M112.346296.
30. Bátkai S, Pacher P, Osei-Hyiaman D, Radaeva S, Liu J, Harvey-White J, Offertáler L, Mackie K, Rudd MA, Bukoski RD, Kunos G. Endocannabinoids acting at cannabinoid-1 receptors regulate cardiovascular function in hypertension. Circulation. 2004;110(14):1996–2002. https://doi.org/10.1161/01.CIR.0000143230.23252.D2.

31. Wagner JA, Járai Z, Bátkai S, Kunos G. Hemodynamic effects of cannabinoids: coronary and cerebral vasodilation mediated by cannabinoid CB(1) receptors. Eur J Pharmacol. 2001a;423(2–3):203–10. https://doi.org/10.1016/s0014-2999(01)01112-8.
32. Wagner JA, Hu K, Bauersachs J, Karcher J, Wiesler M, Goparaju SK, Kunos G, Ertl G. Endogenous cannabinoids mediate hypotension after experimental myocardial infarction. J Am Coll Cardiol. 2001b;38(7):2048–54. https://doi.org/10.1016/s0735-1097(01)01671-0.
33. Marshall NJ, Liang L, Bodkin J, Dessapt-Baradez C, Nandi M, Collot-Teixeira S, Smillie SJ, Lalgi K, Fernandes ES, Gnudi L, Brain SD. A role for TRPV1 in influencing the onset of cardiovascular disease in obesity. Hypertension. 2013;61(1):246–52. https://doi.org/10.1161/HYPERTENSIONAHA.112.201434.
34. Wilson JL, Duan R, El-Marakby A, Alhashim A, Lee DL. Peroxisome proliferator activated receptor-α agonist slows the progression of hypertension, attenuates plasma Interleukin-6 levels and renal inflammatory markers in angiotensin II infused mice. PPAR Res. 2012;2012:645969. https://doi.org/10.1155/2012/645969.
35. Yu Y, Xue BJ, Wei SG, Zhang ZH, Beltz TG, Guo F, Johnson AK, Felder RB. Activation of central PPAR-γ attenuates angiotensin II-induced hypertension. Hypertension. 2015;66(2):403–11. https://doi.org/10.1161/HYPERTENSIONAHA.115.05726.
36. Sexton A, McDonald M, Cayla C, Thiemermann C, Ahluwalia A. 12-lipoxygenase-derived eicosanoids protect against myocardial ischemia/reperfusion injury via activation of neuronal TRPV1. FASEB J. 2007;21(11):2695–703. https://doi.org/10.1096/fj.06-7828com.
37. Maeda N, Osanai T, Kushibiki M, Fujiwara T, Tamura Y, Oowada S, Higuma T, Sasaki S, Yokoyama J, Yoshimachi F, Matsunaga T, Hanada H, Okumura K. Increased serum anandamide level at ruptured plaque site in patients with acute myocardial infarction. Fundam Clin Pharmacol. 2009;23(3):351–7. https://doi.org/10.1111/j.1472-8206.2009.00679.x.
38. Di Filippo C, Rossi F, Rossi S, D'Amico M. Cannabinoid CB2 receptor activation reduces mouse myocardial ischemia-reperfusion injury: involvement of cytokine/chemokines and PMN. J Leukoc Biol. 2004;75(3):453–9. https://doi.org/10.1189/jlb.0703303.
39. Steffens S, Veillard N, Arnaud C, et al. Low dose oral cannabinoid therapy reduces progression of atherosclerosis in mice. Nature. 2005;434:782–6. https://doi.org/10.1038/nature03389.
40. Bátkai S, Rajesh M, Mukhopadhyay P, Haskó G, Liaudet L, Cravatt BF, Csiszár A, Ungvári Z, Pacher P. Decreased age-related cardiac dysfunction, myocardial nitrative stress, inflammatory gene expression, and apoptosis in mice lacking fatty acid amide hydrolase. Am J Physiol Heart Circ Physiol. 2007;293(2):H909–18. https://doi.org/10.1152/ajpheart.00373.2007.
41. Nissen SE, Nicholls SJ, Wolski K, Rodés-Cabau J, Cannon CP, Deanfield JE, Després JP, Kastelein JJ, Steinhubl SR, Kapadia S, Yasin M, Ruzyllo W, Gaudin C, Job B, Hu B, Bhatt DL, Lincoff AM, et al. Effect of rimonabant on progression of atherosclerosis in patients with abdominal obesity and coronary artery disease: the STRADIVARIUS randomized controlled trial. JAMA. 2008;299(13):1547–60. https://doi.org/10.1001/jama.299.13.1547.
42. O'Leary DH, Reuwer AQ, Nissen SE, Després JP, Deanfield JE, Brown MW, Zhou R, Zabbatino SM, Job B, Kastelein JJ, Visseren FL, AUDITOR investigators. Effect of rimonabant on carotid intima-media thickness (CIMT) progression in patients with abdominal obesity and metabolic syndrome: the AUDITOR trial. Heart. 2011;97(14):1143–50. https://doi.org/10.1136/hrt.2011.223446.
43. Rajesh M, Mukhopadhyay P, Haskó G, Liaudet L, Mackie K, Pacher P. Cannabinoid-1 receptor activation induces reactive oxygen species-dependent and -independent mitogen-activated protein kinase activation and cell death in human coronary artery endothelial cells. Br J Pharmacol. 2010;160(3):688–700. https://doi.org/10.1111/j.1476-5381.2010.00712.x.
44. Rajesh M, Mukhopadhyay P, Bátkai S, Haskó G, Liaudet L, Huffman JW, Csiszar A, Ungvari Z, Mackie K, Chatterjee S, Pacher P. CB2-receptor stimulation attenuates TNF-alpha-induced human endothelial cell activation, transendothelial migration of monocytes, and monocyte-endothelial adhesion. Am J Physiol Heart Circ Physiol. 2007;293(4):H2210–8. https://doi.org/10.1152/ajpheart.00688.2007.

45. Zhou Y, Wang X, Guo L, Chen L, Zhang M, Chen X, Li J, Zhang L. TRPV1 activation inhibits phenotypic switching and oxidative stress in vascular smooth muscle cells by upregulating PPARα. Biochem Biophys Res Commun. 2021 Mar 19;545:157–63. https://doi.org/10.1016/j.bbrc.2021.01.072. Epub 2021 Feb 4. PMID: 33550097.
46. Li BH, Yin YW, Liu Y, Pi Y, Guo L, Cao XJ, Gao CY, Zhang LL, Li JC. TRPV1 activation impedes foam cell formation by inducing autophagy in oxLDL-treated vascular smooth muscle cells. Cell Death Dis. 2014;5(4):e1182. https://doi.org/10.1038/cddis.2014.146.
47. Babaev VR, Ishiguro H, Ding L, Yancey PG, Dove DE, Kovacs WJ, Semenkovich CF, Fazio S, Linton MF. Macrophage expression of peroxisome proliferator-activated receptor-alpha reduces atherosclerosis in low-density lipoprotein receptor-deficient mice. Circulation. 2007;116(12):1404–12. https://doi.org/10.1161/CIRCULATIONAHA.106.684704.
48. Reinhard W, Stark K, Neureuther K, Sedlacek K, Fischer M, Baessler A, Weber S, Kaess B, Wiedmann S, Erdmann J, Lieb W, Jeron A, Riegger G, Hengstenberg C. Common polymorphisms in the cannabinoid CB2 receptor gene (CNR2) are not associated with myocardial infarction and cardiovascular risk factors. Int J Mol Med. 2008;22(2):165–74.
49. Auer R, Sidney S, Goff D, Vittinghoff E, Pletcher MJ, Allen NB, Reis JP, Lewis CE, Carr J, Rana JS. Lifetime marijuana use and subclinical atherosclerosis: the coronary artery risk development in young adults (CARDIA) study. Addiction. 2018;113(5):845–56. https://doi.org/10.1111/add.14110.
50. Reis JP, Auer R, Bancks MP, Goff DC Jr, Lewis CE, Pletcher MJ, Rana JS, Shikany JM, Sidney S. Cumulative lifetime marijuana use and incident cardiovascular disease in middle age: the coronary artery risk development in young adults (CARDIA) study. Am J Public Health. 2017;107(4):601–6. https://doi.org/10.2105/AJPH.2017.303654.
51. Desai R, Fong HK, Shah K, Kaur VP, Savani S, Gangani K, Damarlapally N, Goyal H. Rising trends in hospitalizations for cardiovascular events among young cannabis users (18-39 years) without other substance abuse. Medicina (Kaunas). 2019;55(8):438. https://doi.org/10.3390/medicina55080438.
52. Desai R, Shamim S, Patel K, Sadolikar A, Kaur VP, Bhivandkar S, Patel S, Savani S, Mansuri Z, Mahuwala Z. Primary causes of hospitalizations and procedures, predictors of in-hospital mortality, and trends in cardiovascular and cerebrovascular events among recreational marijuana users: a five-year nationwide inpatient assessment in the United States. Cureus. 2018;10(8):e3195. https://doi.org/10.7759/cureus.3195.
53. Johnson-Sasso CP, Tompkins C, Kao DP, Walker LA. Marijuana use and short-term outcomes in patients hospitalized for acute myocardial infarction. PLoS One. 2018;13(7):e0199705. https://doi.org/10.1371/journal.pone.0199705.
54. Liao Y, Bin J, Asakura M, Xuan W, Chen B, Huang Q, Xu D, Ledent C, Takashima S, Kitakaze M. Deficiency of type 1 cannabinoid receptors worsens acute heart failure induced by pressure overload in mice. Eur Heart J. 2012;33(24):3124–33. https://doi.org/10.1093/eurheartj/ehr246.
55. Liao Y, Bin J, Luo T, Zhao H, Ledent C, Asakura M, Xu D, Takashima S, Kitakaze M. CB1 cannabinoid receptor deficiency promotes cardiac remodeling induced by pressure overload in mice. Int J Cardiol. 2013a;167(5):1936–44. https://doi.org/10.1016/j.ijcard.2012.05.033.
56. Liao Y, Bin J, Luo T, Zhao H, Ledent C, Asakura M, Xu D, Takashima S, Kitakaze M. CB1 cannabinoid receptor deficiency promotes cardiac remodeling induced by pressure overload in mice. Int J Cardiol. 2013b;167(5):1936–44. https://doi.org/10.1016/j.ijcard.2012.05.033.
57. Wu L, et al. Marijuana use is associated with better hospital outcome in patients with acute heart failure: a propensity match analysis from national inpatient database. J Card Fail. 2019;25(8):S158.
58. Kristman-Valente AN, Hill KG, Epstein M, Kosterman R, Bailey JA, Steeger CM, Jones TM, Abbott RD, Johnson RM, Walker D, David HJ. The relationship between marijuana and conventional cigarette smoking behavior from early adolescence to adulthood. Prev Sci. 2017;18(4):428–38. https://doi.org/10.1007/s11121-017-0774-4.
59. Rodondi N, Pletcher MJ, Liu K, Hulley SB, Sidney S. Coronary artery risk development in young adults (CARDIA) study. Marijuana use, diet, body mass index, and cardiovascular risk factors (from the CARDIA study). Am J Cardiol. 2006;98:478–84. https://doi.org/10.1016/j.amjcard.2006.03.024.

Respiratory System

6

While one of the most important safety features of cannabis as a medicine is the absence of significant respiratory depression, a feature not shared with opiates and which makes them so dangerous, that does not mean that the endocannabinoid system does not play a role in the respiratory system. In fact, it is involved in both central and peripheral control of respiration as well as control of inflammation in airways and even bronchodilation. Within the lungs themselves, CB1 and CB2 are expressed mainly in immune cells. CB2 receptors are densely located in monocytes and eosinophils and play a role in eosinophil migration, playing a role in potential chronic inflammation. Its presence on dendritic cells, which present foreign antigens to T cells and T cell cells themselves, is decidedly anti-inflammatory though. Both receptors are also present in sensory nerves, reducing neurogenic inflammation, specifically in cholinergic nerves in the lungs [1]. Despite the known ability of the ECS to relax smooth muscle cells in the GI tract, there does not appear evidence of direct action here but rather presynaptic activity, resulting in both vasoconstriction and vasodilation.

6.1 Central Control of Respiration

Central control of respiration is a complicated process with both voluntary and reflexive control, many different inputs are needed to maintain homeostasis, and so at this point in the book, it should come as no surprise that the ECS is extensively involved as well. Reflexive inspiratory drive is initiated by the preBotzinger complex, a nucleus of the ventral lateral medulla. It coordinates with the dorsal respiratory group and with hypoglossal and vagus efferent nerves to set a respiratory rhythm. It also coordinates with the pneumotaxic center of the dorsal region of the pons to transition from inspiration to expiration and from the apneustic center of the lower pons to transition from expiration to inspiration. Microglia here play a not clearly defined role in promoting respiration. Mechanoreceptors in the lungs,

J.-P. H. Dedam et al., *Diseases of the Endocannabinoid System*,
https://doi.org/10.1007/978-3-031-87484-0_6

intercostal muscles, and the diaphragm, as well as chemo- and baroreceptor in the carotid arch, provide feedback via the nucleus tractus solitarius to adjust respiratory rate. Hypoglossal and vagus afferent nerves also provide inputs here to factor in CO_2, O_2 levels, and pH. Another important respiratory center is the retrotrapezoid nucleus that controls active respiration such as with exercise and CO_2 buildup [2].

There are case reports of respiratory depression resulting from the use of synthetic full agonist cannabinoids, known colloquially as spice [3]. These findings are collaborated by studies in rats using similar full agonist of the CB1receptors, resulting in respiratory depression independent of opiate receptor activation [4]. Cannabinoid receptors are present in the preBotzinger complex, and these may be responsible for the preceding finding. Of interest, however, are recent studies showing that activation of CB2 attenuated respiratory depression caused by opiates, although they number far fewer than opiate receptors in the area [5]. CB2 receptors are not present in neural tissues itself but are present in microglial cells located in respiratory centers and are necessary for their function. As noted above, microglia play a poorly defined role in the promotion of respiration, so this may be the mechanism behind the above experimental finding. Also, a peripherally restricted CB1 receptor agonist was used experimentally and in addition to not having any respiratory depression effect (as this affect is centrally mediated), but also attenuated opiate-dependent respiratory depression [6]. Given that there are no reported instances of respiratory depression related to simple cannabis use, and limited number of CB1 receptors in the respiratory centers of the brain stem, it would take an extremely high amount of stimulation, such as acute spice intoxication in a child, to elicit respiratory depression. And the beneficial effects of CB2 receptor activation centrally and CB1 and other receptors peripherally appear to be enough to make respiratory failure nonrealistic due to normal cannabis consumption.

As was seen in the previous chapter on the cardiovascular system, activation of the endocannabinoid system in the nucleus tractus solitarius can cause a baroreflex response. In a separate study, activation of the baroreflex has been shown to alter respiration, prolonging the expiration phase and overall decreasing the respiratory rate [7]. This could represent another point of modulation by the ECS via either a peripheral or central mechanism. CB1 receptors are also present in nerve endings as well as the nuclei of both the vagus and hypoglossal nerves, which relay information of O_2, CO_2, and pH from the periphery to the central control groups, representing yet another unstudied potential area of modification of respiration by the ECS. Two small studies have shown improvement in adult obstructive sleep apnea with dronabinol, a synthetic THC, with improvement in the apnea-hypopnea index [8, 9]. The K+ channel receptor TWIK-related acid-sensitive potassium channel (TASK-1) is known to have a role in pathophysiology of OSA, and a novel antagonist has been shown to have benefit in treating the condition [10]. AEA is also a known blocker of the TASK-1 receptor and is a possible explanation for the beneficial effects of dronabinol on OSA [11]. Of course, as seen in the chapter of metabolism and obesity, overactivation of the ECS can drive metabolic syndrome, an independent risk factor for the development of OSA. Levels of circulating AEA are known to be increased in OSA, and this is likely due to metabolic syndrome and not a compensatory mechanism against hypoxia [12]. AEA is also capable of activating TRPV1 receptors on afferent nerve fibers in the carotid body that project again to the nucleus tractus solitarius. Increases in temperature such as during exercise, another physiologic condition where AEA is known to be increased, cause activation of TRPV1 by AEA and results in increased respiration. While baroreflex is also activated at the same time, experiments in rats show that chemoreflex to increase heat is dominant and the respiration rate increases [13]. Activation of CB1 has also been shown in studies on rats with OSA to decrease apnea; however, the mechanism and site of this effect are unknown and require further investigation [14].

6.2 ECS and Bronchodilation

Moving on the direct effects on lung tissues themselves, human trials from the 1970s did demonstrate bronchodilation by oral and aerosolized but not smoked THC in healthy and asthmatic subjects; however, some subjects did develop bronchoconstriction instead [15, 16]. Control of bronchodilation by the ECS appears to be conditional. When bronchi are constricted such as in capsaicin-evoked experimental models, AEA prompts bronchodilation. However, if the airways were already relaxed by removing vagal inputs with atropine, AEA could cause bronchoconstriction, both effects via CB1 receptor [17].

6.3 ECS and Lung Inflammatory

Another role of the ECS within the lungs themselves is their modulation on inflammation. The anti-inflammatory effects of the ECS are well documented in other systems but not the skin, the GI tract, and CV system, and this is also true for lung tissues. Studies in mice models using bacterial LPS to induce inflammation found decreased TNF and neutrophil migration [18]. ECS activation had a similar effect on macrophages [19]. The ECS has a similar effect on dendritic cells, prominent antigen resenting cells, both decreasing their maturation and migration via CB2 receptor activation [20, 21] in addition to well-studied anti-inflammatory changes in T cells. This is potentially balanced by findings that the endocannabinoid 2-AG can actually actively attract eosinophils, who are players in the inflammatory process of chronic asthma [22]. The clinical significance of the above changes in immune response is mixed. Mice studies confirmed that selective activation of CB2 receptor could worsen allergic airway diseases in mice models [23]. Under different physiologic conditions, activation of CB2 receptor and subsequent decrease in inflammation decreased mortality in mice with acute lung injury induced by paraquat [24]. Another study in mice, using very high doses of THC, found increased mortality in mice due to decreased cytokine response in sublethal infections of *Legionella* [25]. However, other sepsis models show clear benefit to survival in mice as well as reduced lung injury when treated with CB2 selective agonist due in part to decreased circulating IL-6 and neutrophil recruitment [26]. Chronic cannabis use in humans has been tied to increased chronic bronchitis without increased risk of COPD or emphysema [27]; however, these changes are also certainly due to smoking.

6.4 ECS and Asthma

There is a correlation between atopy and asthma in patients, and as we saw in the chapter on skin, treating an underactive ECS did improve atopy. It is possible that the chronic inflammation of asthma may have a similar endocannabinoid deficiency component. A study in guinea pigs found that administration of a synthetic cannabinoid that activated both CB1 and CB2 was capable of preventing an antigen-induced, asthma-like reaction, with significant decline in anti-inflammatory markers present in bronchial lavage [28]. Inhibition of MAG-L and boosting 2-AG levels, also in guinea pigs, decreased inflammation such as improved airway reactivity [29]. When considering other receptors in the broader endocannabinoid family, we again get a mixed picture, with TRPV1 being associated with pathogenesis of asthma and PPAR's having a potentially beneficial effect such as decreasing inflammation and deceasing airway hyperresponsiveness [30, 31]. CBD administered to asthma disease-modeled mice improved both inflammation and airway hyperreactiveness, and given its low affinity for CB1 and CB2 receptors, this may be mediated via PPAR [32].

At nearly level, the effects of the ECS produce some beneficial and some detrimental effects. One thing is clear though, further study is warranted to better understand its role and potential use in treating a wide variety of pulmonary conditions.

References

1. Bozkurt TE. Endocannabinoid system in the airways. Molecules. 2019;24(24):4626. https://doi.org/10.3390/molecules24244626.
2. Wiese BM, Alvarez Reyes A, Vanderah TW, Largent-Milnes TM. The endocannabinoid system and breathing. Front Neurosci. 2023;17:1126004. https://doi.org/10.3389/fnins.2023.1126004.
3. Jinwala FN, Gupta M. Synthetic cannabis and respiratory depression. J Child Adolesc Psychopharmacol. 2012;22(6):459–62. https://doi.org/10.1089/cap.2011.0122.
4. Pfitzer T, Niederhoffer N, Szabo B. Central effects of the cannabinoid receptor agonist WIN55212-2 on respiratory and cardiovascular regulation in anaesthetised rats. Br J Pharmacol. 2004;142(6):943–52. https://doi.org/10.1038/sj.bjp.0705874.
5. Wiese BM, Liktor-Busa E, Levine A, Couture SA, Nikas SP, Ji L, Liu Y, Mackie K, Makriyannis A, Largent-Milnes TM, Vanderah TW. Cannabinoid-2 agonism with AM2301 mitigates morphine-induced respiratory depression. Cannabis Cannabinoid Res. 2021;6(5):401–12. https://doi.org/10.1089/can.2020.0076.
6. Wiese BM, Liktor-Busa E, Couture SA, Nikas SP, Ji L, Liu Y, Makriyannis A, Spigelman I, Vanderah TW, Largent-Milnes TM. Brain penetrant, but not peripherally restricted, synthetic cannabinoid 1 receptor agonists promote morphine-mediated respiratory depression. Cannabis Cannabinoid Res. 2022;7(5):621–7. https://doi.org/10.1089/can.2021.0090.

7. Baekey DM, Molkov YI, Paton JF, Rybak IA, Dick TE. Effect of baroreceptor stimulation on the respiratory pattern: insights into respiratory-sympathetic interactions. Respir Physiol Neurobiol. 2010;174(1–2):135–45. https://doi.org/10.1016/j.resp.2010.09.006.
8. Prasad B, Radulovacki MG, Carley DW. Proof of concept trial of dronabinol in obstructive sleep apnea. Front Psych. 2013;4:1. https://doi.org/10.3389/fpsyt.2013.00001.
9. Carley DW, Prasad B, Reid KJ, Malkani R, Attarian H, Abbott SM, Vern B, Xie H, Yuan C, Zee PC. Pharmacotherapy of apnea by cannabimimetic enhancement, the PACE clinical trial: effects of dronabinol in obstructive sleep apnea. Sleep. 2018;41(1):zsx184. https://doi.org/10.1093/sleep/zsx184.
10. Osman AM, Toson B, Naik G, Mukherjee S, Delbeck M, Hahn M, Mueller T, Weimann G, Eckert DJ. A novel TASK channel antagonist nasal spray reduces sleep apnea severity in physiological responders: a randomized, blinded, trial. Am J Physiol Heart Circ Physiol. 2024;326:H715. https://doi.org/10.1152/ajpheart.00541.2023.
11. Maingret F, Patel AJ, Lazdunski M, Honoré E. The endocannabinoid anandamide is a direct and selective blocker of the background K(+) channel TASK-1. EMBO J. 2001;20(1–2):47–54. https://doi.org/10.1093/emboj/20.1.47.
12. Wang X, Yu Q, Yue H, Zhang J, Zeng S, Cui F. Circulating endocannabinoids and insulin resistance in patients with obstructive sleep apnea. Biomed Res Int. 2016;2016:1. https://doi.org/10.1155/2016/9782031.
13. Roy A, Mandadi S, Fiamma MN, Rodikova E, Ferguson EV, Whelan PJ, Wilson RJ. Anandamide modulates carotid sinus nerve afferent activity via TRPV1 receptors increasing responses to heat. J Appl Physiol (1985). 2012 Jan;112(1):212–24. https://doi.org/10.1152/japplphysiol.01303.2010. Epub 2011 Sep 8. Erratum in: J Appl Physiol. 2013;Feb15;114(4):521. PMID: 21903882.
14. Calik MW, Radulovacki M, Carley DW. Intranodose ganglion injections of dronabinol attenuate serotonin-induced apnea in Sprague-Dawley rat. Respir Physiol Neurobiol. 2014;Jan 1;190:20–4. https://doi.org/10.1016/j.resp.2013.10.001. Epub 2013 Oct 11. PMID: 24121138; PMCID: PMC3880550.
15. Abboud RT, Sanders HD. Effect of oral administration of delta-tetrahydrocannabinol on airway mechanics in normal and asthmatic subjects. Chest. 1976;70(4):480–5. https://doi.org/10.1378/chest.70.4.480.
16. Tashkin DP, Reiss S, Shapiro BJ, Calvarese B, Olsen JL, Lodge JW. Bronchial effects of aerosolized delta 9-tetrahydrocannabinol in healthy and asthmatic subjects. Am Rev Respir Dis. 1977;115(1):57–65. https://doi.org/10.1164/arrd.1977.115.1.57.
17. Calignano A, Kátona I, Désarnaud F, Giuffrida A, La Rana G, Mackie K, Freund TF, Piomelli D. Bidirectional control of airway responsiveness by endogenous cannabinoids. Nature. 2000;408(6808):96–101. https://doi.org/10.1038/35040576.
18. Berdyshev E, Boichot E, Corbel M, Germain N, Lagente V. Effects of cannabinoid receptor ligands on LPS-induced pulmonary inflammation in mice. Life Sci. 1998;63(8):PL125. https://doi.org/10.1016/s0024-3205(98)00324-5.
19. Zheng ZM, Specter SC. Delta-9-tetrahydrocannabinol suppresses tumor necrosis factor alpha maturation and secretion but not its transcription in mouse macrophages. Int J Immunopharmacol. 1996;18(1):53–68. https://doi.org/10.1016/0192-0561(95)00107-7.
20. Karmaus PW, Chen W, Crawford R, Kaplan BL, Kaminski NE. Δ9-tetrahydrocannabinol impairs the inflammatory response to influenza infection: role of antigen-presenting cells and the cannabinoid receptors 1 and 2. Toxicol Sci. 2013;131(2):419–33. https://doi.org/10.1093/toxsci/kfs315.
21. Adhikary S, Kocieda VP, Yen JH, Tuma RF, Ganea D. Signaling through cannabinoid receptor 2 suppresses murine dendritic cell migration by inhibiting matrix metalloproteinase 9 expression. Blood. 2012;120(18):3741–9. https://doi.org/10.1182/blood-2012-06-435362.
22. Larose MC, Turcotte C, Chouinard F, Ferland C, Martin C, Provost V, Laviolette M, Flamand N. Mechanisms of human eosinophil migration induced by the combination of IL-5 and the endocannabinoid 2-arachidonoyl-glycerol. J Allergy Clin Immunol. 2014;133(5):1480–2. https://doi.org/10.1016/j.jaci.2013.12.1081.

23. Frei RB, Luschnig P, Parzmair GP, Peinhaupt M, Schranz S, Fauland A, Wheelock CE, Heinemann A, Sturm EM. Cannabinoid receptor 2 augments eosinophil responsiveness and aggravates allergen-induced pulmonary inflammation in mice. Allergy. 2016;71(7):944–56. https://doi.org/10.1111/all.12858.
24. Liu Z, Wang Y, Zhao H, Zheng Q, Xiao L, Zhao M. CB2 receptor activation ameliorates the proinflammatory activity in acute lung injury induced by paraquat. Biomed Res Int. 2014;2014:971750. https://doi.org/10.1155/2014/971750.
25. Smith MS, Yamamoto Y, Newton C, Friedman H, Klein T. Psychoactive cannabinoids increase mortality and alter acute phase cytokine responses in mice sublethally infected with Legionella pneumophila. Proc Soc Exp Biol Med. 1997;214(1):69–75. https://doi.org/10.3181/00379727-214-44071.
26. Tschöp J, Kasten KR, Nogueiras R, Goetzman HS, Cave CM, England LG, Dattilo J, Lentsch AB, Tschöp MH, Caldwell CC. The cannabinoid receptor 2 is critical for the host response to sepsis. J Immunol. 2009;183(1):499–505. https://doi.org/10.4049/jimmunol.0900203.
27. Gracie K, Hancox RJ. Cannabis use disorder and the lungs. Addiction. 2021;116(1):182–90. https://doi.org/10.1111/add.15075.
28. Giannini L, Nistri S, Mastroianni R, Cinci L, Vannacci A, Mariottini C, Passani MB, Mannaioni PF, Bani D, Masini E. Activation of cannabinoid receptors prevents antigen-induced asthma-like reaction in Guinea pigs. J Cell Mol Med. 2008;12(6A):2381–94. https://doi.org/10.1111/j.1582-4934.2008.00258.x.
29. Abohalaka R, Karaman Y, Recber T, Onder SC, Nemutlu E, Bozkurt TE. Endocannabinoid metabolism inhibition ameliorates ovalbumin-induced allergic airway inflammation and hyperreactivity in Guinea pigs. Life Sci. 2022;306:120808. https://doi.org/10.1016/j.lfs.2022.120808.
30. Reyes-García J, Carbajal-García A, Montaño LM. Transient receptor potential cation channel subfamily V (TRPV) and its importance in asthma. Eur J Pharmacol. 2022;915:174692. https://doi.org/10.1016/j.ejphar.2021.174692.
31. Kytikova OY, Perelman JM, Novgorodtseva TP, Denisenko YK, Kolosov VP, Antonyuk MV, Gvozdenko TA. Peroxisome proliferator-activated receptors as a therapeutic target in asthma. PPAR Res. 2020;2020:8906968. https://doi.org/10.1155/2020/8906968.
32. Vuolo F, Abreu SC, Michels M, Xisto DG, Blanco NG, Hallak JE, et al. Cannabidiol reduces airway inflammation and fibrosis in experimental allergic asthma. Eur J Pharmacol. 2019;843:251–9.

Kidney and Urinary System

7

Aside from its most immediate roles of removing toxins from blood and maintaining homeostasis of electrolyte levels and pH, the kidney also regulates blood pressure, produces erythropoietin, and activates vitamin D. The role of the endocannabinoid system in many of these functions is still be elucidated. We will present the research as it stands and discusses the role of ECS dysfunction in renal disease. CB1 receptor was overexpressed in podocytes in diabetic nephropathy, but that blocking its activity in those cells prevented further nephropathy similar to the beneficial effects of angiotensin receptor-blocking drugs like losartan.

7.1 Distribution of the ECS in the Kidney

As the anatomy of the kidney is somewhat complex, we will briefly review its major components and its functions. The kidney is made up of two major parts: the cortex and the medulla. The medulla forms numerous pyramids that extend in the renal calyx, where urine is collected and is passed through the ureters down into the bladder. The basic physiologic unit of the kidney is the nephron, which extends from the cortex into the medulla. The nephron consists of the glomeruli, surrounded by Bowman's capsule and podocytes, the proximal tubule, the loop of Henle, and the distal convoluted tubule and collecting ducts. Renal blood flow enters the kidneys via the renal arteries and then branches into the interlobar arteries and arcuate arteries and then into the efferent arterioles, which enter into the glomeruli forming capillaries and then exit via the afferent arterioles. Vasconstrition of the afferent and efferent arterioles can significantly change filtration rates. Blood is filtered in the glomeruli and this filtrate then passes into the proximal tubules where the majority of reabsorption, both passive and active, of electrolytes, glucose, water, and other substances is achieved. Only approximately 1% of the filtrate will pass through the nephron's various tubes to the collecting system to form urine. Waste products such as ammonia and uric acid are secreted mostly in the distal tubule. Resorption of

J.-P. H. Dedam et al., *Diseases of the Endocannabinoid System*,
https://doi.org/10.1007/978-3-031-87484-0_7

water is controlled by antidiuretic hormone and sodium and potassium by aldosterone [1]. CB1 receptor has already been found in the proximal convoluted tubules, distal tubules, and intercalated cells of the collecting duct, as well as in the afferent and efferent arterioles, glomeruli, and the loop of Henle. CB2 receptor, in addition to immune cells present in the kidney, has also been found in podocytes, proximal tubule cells, and mesangial cells of the glomeruli. Both AEA and 2-AG are present in equal levels in the cortex with increased expression of FAAH compared to the rest of the kidney. AEA is more concentrated in the medulla than 2-AG, which has similar levels in both parts of the kidney. AEA is also found in endothelial and mesangial cells [2].

7.2 Role of ECS in Diabetic Nephropathy

In rat studies, administration of AEA systemically resulted in vasodilation of the efferent and afferent arterioles with a decrease in GFR as well as an increase in renal blood flow without affecting MAP. This was a dose-mediated response with lower doses opening only the efferent arterioles and higher doses opening both. This effect was blocked by a nonselective cannabinoid receptor blocker [3]. This vasodilation was mediated by CB1 receptor-activated production of NO [4]. Both elevation in blood glucose and increase in renin-aldosterone signaling can lead to diabetic nephropathy, specifically with destruction of podocytes and disruption of filtration at Bowman's membrane. Studies in rats not only showed that CB1 receptor was overexpressed in the podocytes in diabetic nephropathy, but that blocking its activity in those cells prevented further nephropathy similar to the beneficial effects of angiotensin receptor-blocking drugs like losartan [5]. Given that overactivation of the ECS is present in metabolic syndrome systemically, with increases in circulating AEA levels, the ECS may actually accelerate nephropathy due to diabetes and possibly HTN as well. In separate studies, activation of CB1 receptor in mice also increased protein secretion and resulted in proteinuria, this time by increasing the expression of VEGF [6].

Moving on to tubules, activation of CB1R by AEA in the loop of Henle prevented normal Na ion pump function. This could potentially influence salt balance in the blood [7]. Intramedullary injection of AEA in a separate study increased urine volume as well as sodium and potassium excretion without a change in blood pressure, an effect that is blocked by COX-2 inhibitors, indicating that it is likely a prostaglandin metabolite rather than AEA itself that is having this effect [8]. Administration of THC in rats has also been shown to produce similar diuretic effects mediated by CB1 receptor [9]. As was seen with podocytes above, elevation on blood glucose also causes increased CB1 receptor expression in the proximal tubule. This increased activity results in the activation of mTORC1, a sensor/signaling pathway involved in carbohydrate management, which increased GLUT2 receptors. These receptors reabsorb glucose out of the filtrate and then into the extracellular space, increasing their concentrations there, again possibly promoting the development of diabetic kidney disease [10]. A small study in humans showed increased levels of CB1 expression in patients with IgA nephropathy, diabetic nephropathy, and acute interstitial nephritis, mainly in endothelial cells but also in tubules, interstitial cells, and podocytes in diabetic nephropathy [11]. TRPV1 activation appears to play a role in decreasing renal inflammation as TRPV1 knockout rats with salt-induced HTN were noted to suffer from considerably more inflammation and subsequent decline in renal function [12].

7.3 Role of ECS in Fibrosis After Acute Urinary Obstruction

Experimental models of fibrosis in mice have shown that CB1 receptor expression in the renal tissue is also dramatically increased after unilateral obstruction of the ureter. This leads to increased fibrosis with increased expression of CB1 in myofibroblasts [11]. CB2 receptor activation under the same experimental conditions decreased fibrosis, although the exact mechanism is unknown [13]. A new CB2 receptor agonist is even being pursued as a treatment for inflammatory renal disease with evidence of reduced fibrosis in the same unilateral obstruction of the ureter experimental model [14]. As seen in other organ systems, the inflammatory properties of CB2 receptor activation may represent a number of potential treatment applications.

7.4 Role of ECS in AKI

In the case of acute kidney injury (AKI) due to ischemic insults, both CB1 and CB2 have shown benefit in preventing injury when administered prior to ischemia, similar to other studies of cardiac and cerebral ischemia [15]. Other studies using the same model found increased levels of 2-AG in the kidney, and these levels correlated to improved renal function [16]. However, in experimental models of mice looking at cisplatin-induced nephropathy, CB1 played a negative role in the development of AKI via the MAPK signaling pathway, while CB2 receptor activation continued to be protective [17, 18]. CB2 receptor activation has also been found to be protective in AKI due to sepsis, reducing mortality by decreasing circulating IL-6 levels and neutrophil recruitment [19]. All of these studies again demonstrate the potential pharmaceutical targeting of CB2 receptor in the treatment and prevention of AKI and nephropathy, especially given its lack of psychoactive side effects. Synthetic cannabinoids used recreationally, typically close to full CB1 receptor agonists, have also been associated with the development of AKI with acute tubular damage and acute inter interstitial nephritis found on biopsy. The mechanism of action for the injury is unknown, although volume depletion is component [20].

7.5 Autosomal Dominant Polycystic Kidney Disease (ADPKD)

In a study comparing 100 patients with autosomal dominant polycystic kidney disease (ADPKD) with normal controls, decreased levels of endocannabinoids were detected. This is unexpected given that ADPKD results in significant interstitial inflammation, which is in turn linked to increased cyst progression. Patients with higher a-AG levels were also associated with better renal function. This could represent some degree of ECS deficiency. Another theory put forward by the author was an overactive MAGL enzyme. This would have two effects: reducing the amount of available 2-AG and increasing pro-inflammatory 2-AG metabolites like arachidonic acid [21]. Endocannabinoid levels have also been looked at in patients with ESRD on hemodialysis, with elevated levels of 2-AG found in patients versus controls, with 2-AG level correlating with markers of metabolic disease [22]. In a separate study by the same investigators, elevations in 2-AG were associated with decreased all-cause mortality. AEA levels did not appear to correlate to mortality; it was associated with higher levels of IL-6 [23].

7.6 ECS in Hypertensive Kidney Disease

Central control of blood pressure was previously covered in the chapter on the cardiovascular system, with an overall decrease in systemic BP as the dominant effect of the ECS. However, it was noted that the ECS is necessary for the renin-angiotensin system to elevate blood pressure centrally. Locally in the kidney, hypertension causes renal oxidative stress, which can then cause kidney damage. Reactive oxygen species created under these conditions cause significant inflammation as well further worsening of the renal function. It has been purposed that an imbalance between reactive oxygen species and antioxidant protection can spontaneously lead to the development of HTN itself. In a recent study, rats either with spontaneous HTN or with HTN induced by salt intake were treated with a blocker of the enzyme FAAH to boost AEA levels. This resulted in decreased reactive oxygen species and decreased inflammation, particularly in rats with salt-induced HTN with increases in the expression of CB1, CB2, and TRPV1 in the kidney [24]. Whereas overactivation in the context of metabolic disease may drive HTN development systemically, locally in the kidney, it appears to have the opposite effect. A localized ECS deficiency could play a role in the development and progression of HTN at the level of the kidney, despite the fact that isolated CB1 activation worsens nephropathy, especially in diabetics.

7.7 ECS in the Urinary Collecting System

Once the urine leaves the kidney, it travels down the ureter to the bladder. CB1 and TRPV1 receptors are present throughout the bladder tissue, with CB2 receptor found in the urothelium and detrusor muscles including nerves that control both sensory and motor functions in the bladder. CB1 is active both presynaptically and postsynaptically in inhibiting bladder contraction, while CB2 is active only postsynaptically [25]. CB1 activation in bladder resulted in inhibition of nerve-mediated contractions in mice [26]. Activation of both receptors has actually been shown to decrease contractions in the human bladder [27]. Numerous studies have also shown the effect of both receptors on modulating spinal afferents from the bladder, with CB1 tending to decrease micturition and increasing bladder capacity by suppressing afferent signals and CB2 decreasing hyperreflexia of the bladder, otherwise known as bladder spasm [28]. This is all in line with findings in numerous studies on treatment of bladder symptoms in MS patients with cannabinoids. Sativex, a sublingual spray containing 50/50 THC/CBD ratio, has been shown to decrease episodes of nocturia, urgency, and incontinence with improvements in pain, spasm, and quality of sleep [29]. In studies of chemical-induced cystitis in rodents, AEA's effect on pain and spasm was also mediated by TPRV1 receptor present in the bladder tissue [30].

7.8 Interstitial Cystitis

Cystitis, whether acute or chronic, is also a condition characterized by significant inflammation, as the name implies. The known anti-inflammatory effects of CB2 activation also resulted in improved inflammatory markers but also symptoms of frequency and pain [31]. Similar effects have also been observed with administration of PEA in rats [32]. Interstitial cystitis is a chronic, often idiopathic, and therefore difficult to diagnose condition that causes severe bladder symptoms. In a study of 75 patients, in an attempt to find biomarkers to differentiate interstitial cystitis from overactive bladder syndrome, anandamide levels were found to be significantly lower in patients with interstitial cystitis [33]. This could indicate endocannabinoid deficiency playing a role in the pathogenesis of this condition, especially in light of most acute cystitis experiments above reporting increased ECS expression. Animal models of interstitial cystitis have showed better efficacy of the CB2 agonist beta-caryophyllene than DMSO, the only FDA-approved treatment for the condition [34]. Given its important role in modulating pain, ECS deficiency could also be expected in syndromes of increased lower urinary tract hyperalgesia with preliminary studies on rats showing that boosting the ECS by inhibiting FAAH is an effective strategy for its treatment [35].

References

1. Wallace MA. Anatomy and physiology of the kidney. AORN J. 1998;68(5):800.
2. Chua JT, Argueta DA, DiPatrizio NV, Kovesdy CP, Vaziri ND, Kalantar-Zadeh K, Moradi H. Endocannabinoid system and the kidneys: from renal physiology to injury and disease. Cannabis Cannabinoid Res. 2019;4(1):10–20. https://doi.org/10.1089/can.2018.0060.
3. Koura Y, Ichihara A, Tada Y, Kaneshiro Y, Okada H, Temm CJ, Hayashi M, Saruta T. Anandamide decreases glomerular filtration rate through predominant vasodilation of efferent arterioles in rat kidneys. J Am Soc Nephrol. 2004;15(6):1488–94. https://doi.org/10.1097/01.asn.0000130561.82631.bc.
4. Deutsch DG, Goligorsky MS, Schmid PC, Krebsbach RJ, Schmid HH, Das SK, Dey SK, Arreaza G, Thorup C, Stefano G, Moore LC. Production and physiological actions of anandamide in the vasculature of the rat kidney. J Clin Invest. 1997;100(6):1538–46. https://doi.org/10.1172/JCI119677.

5. Jourdan T, Szanda G, Rosenberg AZ, Tam J, Earley BJ, Godlewski G, Cinar R, Liu Z, Liu J, Ju C, Pacher P, Kunos G. Overactive cannabinoid 1 receptor in podocytes drives type 2 diabetic nephropathy. Proc Natl Acad Sci USA. 2014;111(50):E5420–8. https://doi.org/10.1073/pnas.1419901111.
6. Hsu YC, Lei CC, Shih YH, Ho C, Lin CL. Induction of proteinuria by cannabinoid receptors 1 signaling activation in CB1 transgenic mice. Am J Med Sci. 2015;Feb;349(2):162–8. https://doi.org/10.1097/MAJ.0000000000000352. PMID: 25474224.
7. Silva GB, Atchison DK, Juncos LI, García NH. Anandamide inhibits transport-related oxygen consumption in the loop of Henle by activating CB1 receptors. Am J Physiol Renal Physiol. 2013;304(4):F376–81. https://doi.org/10.1152/ajprenal.00239.2012.
8. Ritter JK, Li C, Xia M, Poklis JL, Lichtman AH, Abdullah RA, Dewey WL, Li PL. Production and actions of the anandamide metabolite prostamide E2 in the renal medulla. J Pharmacol Exp Ther. 2012;342(3):770–9. https://doi.org/10.1124/jpet.112.196451.
9. Paronis CA, Thakur GA, Bajaj S, Nikas SP, Vemuri VK, Makriyannis A, Bergman J. Diuretic effects of cannabinoids. J Pharmacol Exp Ther. 2013;344(1):8–14. https://doi.org/10.1124/jpet.112.199331.
10. Hinden L, Ahmad M, Hamad S, Nemirovski A, Szanda G, Glasmacher S, Kogot-Levin A, Abramovitch R, Thorens B, Gertsch J, Leibowitz G, Tam J. Opposite physiological and pathological mTORC1-mediated roles of the CB1 receptor in regulating renal tubular function. Nat Commun. 2022;13(1):1783. https://doi.org/10.1038/s41467-022-29124-8.
11. Lecru L, Desterke C, Grassin-Delyle S, Chatziantoniou C, Vandermeersch S, Devocelle A, Vernochet A, Ivanovski N, Ledent C, Ferlicot S, Dalia M, Saïd M, Beaudreuil S, Charpentier B, Vazquez A, Giron-Michel J, Azzarone B, Durrbach A, François H. Cannabinoid receptor 1 is a major mediator of renal fibrosis. Kidney Int. 2015;88(1):72–84. https://doi.org/10.1038/ki.2015.63.
12. Wang Y, Wang DH. Aggravated renal inflammatory responses in TRPV1 gene knockout mice subjected to DOCA-salt hypertension. Am J Physiol Renal Physiol. 2009;297(6):F1550–9. https://doi.org/10.1152/ajprenal.00012.2009.
13. El Tohamy M, Adel M, Rashad El-Menabawy F, Gad GEM, El-Gamal R, El Serougy H. Role of cannabinoid type 2 receptor activation in renal fibrosis induced by unilateral ureteric obstruction in rats. Rep Biochem Mol Biol. 2023;12(1):59–73. https://doi.org/10.52547/rbmb.12.1.59.
14. Nettekoven M, Adam JM, Bendels S, Bissantz C, Fingerle J, Grether U, Grüner S, Guba W, Kimbara A, Ottaviani G, Püllmann B, Rogers-Evans M, Röver S, Rothenhäusler B, Schmitt S, Schuler F, Schulz-Gasch T, Ullmer C. Novel Triazolopyrimidine-derived cannabinoid receptor 2 agonists as potential treatment for inflammatory kidney diseases. ChemMedChem. 2016;11(2):179–89. https://doi.org/10.1002/cmdc.201500218.
15. Feizi A, Jafari MR, Hamedivafa F, Tabrizian P, Djahanguiri B. The preventive effect of cannabinoids on reperfusion-induced ischemia of mouse kidney. Exp Toxicol Pathol. 2008;60(4–5):405–10. https://doi.org/10.1016/j.etp.2008.04.006.
16. Moradi H, Oveisi F, Khanifar E, Moreno-Sanz G, Vaziri ND, Piomelli D. Increased renal 2-Arachidonoylglycerol level is associated with improved renal function in a mouse model of acute kidney injury. Cannabis Cannabinoid Res. 2016;1(1):218–28. https://doi.org/10.1089/can.2016.0013.
17. Mukhopadhyay P, Pan H, Rajesh M, Bátkai S, Patel V, Harvey-White J, Mukhopadhyay B, Haskó G, Gao B, Mackie K, Pacher P. CB1 cannabinoid receptors promote oxidative/nitrosative stress, inflammation and cell death in a murine nephropathy model. Br J Pharmacol. 2010;160(3):657–68. https://doi.org/10.1111/j.1476-5381.2010.00769.x.
18. Mukhopadhyay P, Baggelaar M, Erdelyi K, Cao Z, Cinar R, Fezza F, Ignatowska-Janlowska B, Wilkerson J, van Gils N, Hansen T, Ruben M, Soethoudt M, Heitman L, Kunos G, Maccarrone M, Lichtman A, Pacher P, Van der Stelt M. The novel, orally available and peripherally restricted selective cannabinoid CB2 receptor agonist LEI-101 prevents cisplatin-induced nephrotoxicity. Br J Pharmacol. 2016;173(3):446–58. https://doi.org/10.1111/bph.13338.
19. Tschöp J, Kasten KR, Nogueiras R, Goetzman HS, Cave CM, England LG, Dattilo J, Lentsch AB, Tschöp MH, Caldwell CC. The cannabinoid receptor 2 is critical for the host response to sepsis. J Immunol. 2009;183(1):499–505. https://doi.org/10.4049/jimmunol.0900203.

20. D'Errico S, Zanon M, Radaelli D, Concato M, Padovano M, Scopetti M, Frati P, Fineschi V. Acute kidney injury (AKI) in young synthetic cannabinoids abusers. Biomedicines. 2022;10(8):1936. https://doi.org/10.3390/biomedicines10081936.
21. Klawitter J, Sempio C, Jackson MJ, Smith PH, Hopp K, Chonchol M, Gitomer BY, Christians U, Klawitter J. Endocannabinoid system in polycystic kidney disease. Am J Nephrol. 2022;53(4):264–72. https://doi.org/10.1159/000522113.
22. Moradi H, Park C, Igarashi M, Streja E, Argueta DA, Soohoo M, Daglian J, You AS, Rhee CM, Kashyap ML, DiPatrizio NV, Vaziri ND, Kalantar-Zadeh K, Piomelli D. Serum endocannabinoid levels in patients with end-stage renal disease. J Endocr Soc. 2019;3(10):1869–80. https://doi.org/10.1210/js.2019-00242.
23. Moradi H, Park C, Streja E, Argueta DA, DiPatrizio NV, You AS, Rhee CM, Vaziri ND, Kalantar-Zadeh K, Piomelli D. Circulating endocannabinoids and mortality in hemodialysis patients. Am J Nephrol. 2020;51(2):86–95. https://doi.org/10.1159/000505444.
24. Biernacki M, Ambrożewicz E, Gęgotek A, Toczek M, Bielawska K, Skrzydlewska E. Redox system and phospholipid metabolism in the kidney of hypertensive rats after FAAH inhibitor URB597 administration. Redox Biol. 2018;15:41–50. https://doi.org/10.1016/j.redox.2017.11.022.
25. Bakali E, Elliott RA, Taylor AH, Willets J, Konje JC, Tincello DG. Distribution and function of the endocannabinoid system in the rat and human bladder. Int Urogynecol J. 2013;24(5):855–63. https://doi.org/10.1007/s00192-012-1954-1.
26. Pertwee RG, Fernando SR. Evidence for the presence of cannabinoid CB1 receptors in mouse urinary bladder. Br J Pharmacol. 1996;118(8):2053–8. https://doi.org/10.1111/j.1476-5381.1996.tb15643.x.
27. Tyagi V, Philips BJ, Su R, Smaldone MC, Erickson VL, Chancellor MB, Yoshimura N, Tyagi P. Differential expression of functional cannabinoid receptors in human bladder detrusor and urothelium. J Urol. 2009;181(4):1932–8. https://doi.org/10.1016/j.juro.2008.11.078.
28. Hedlund P. Cannabinoids and the endocannabinoid system in lower urinary tract function and dysfunction. Neurourol Urodyn. 2014;Jan;33(1):46–53. https://doi.org/10.1002/nau.22442. PMID: 24285567.
29. Brady CM, DasGupta R, Dalton C, Wiseman OJ, Berkley KJ, Fowler CJ. An open-label pilot study of cannabis-based extracts for bladder dysfunction in advanced multiple sclerosis. Mult Scler. 2004;10(4):425–33. https://doi.org/10.1191/1352458504ms1063oa.
30. Dinis P, Charrua A, Avelino A, Yaqoob M, Bevan S, Nagy I, Cruz F. Anandamide-evoked activation of vanilloid receptor 1 contributes to the development of bladder hyperreflexia and nociceptive transmission to spinal dorsal horn neurons in cystitis. J Neurosci. 2004;24(50):11253–63. https://doi.org/10.1523/JNEUROSCI.2657-04.2004.
31. Wang ZY, Wang P, Bjorling DE. Treatment with a cannabinoid receptor 2 agonist decreases severity of established cystitis. J Urol. 2014;191(4):1153–8. https://doi.org/10.1016/j.juro.2013.10.102.
32. Pessina F, Capasso R, Borrelli F, Aveta T, Buono L, Valacchi G, Fiorenzani P, Di Marzo V, Orlando P, Izzo AA. Protective effect of palmitoylethanolamide in a rat model of cystitis. J Urol. 2015;193(4):1401–8. https://doi.org/10.1016/j.juro.2014.11.083.
33. Torimoto K, Ueda T, Gotoh D, Kano K, Miyake M, Nakai Y, Hori S, Morizawa Y, Onishi K, Shimizu T, Tomizawa M, Aoki J, Fujimoto K. Serum anandamide and lipids associated with linoleic acid can distinguish interstitial cystitis/bladder pain syndrome from overactive bladder: An exploratory study. Low Urin Tract Symptoms. 2023;Nov;15(6):238–246. https://doi.org/10.1111/luts.12501. Epub 2023 Sep 8. PMID: 37688290.
34. Berger G, Arora N, Burkovskiy I, Xia Y, Chinnadurai A, Westhofen R, Hagn G, Cox A, Kelly M, Zhou J, Lehmann C. Experimental cannabinoid 2 receptor activation by phyto-derived and synthetic cannabinoid ligands in LPS-induced interstitial cystitis in mice. Molecules. 2019;24(23):4239. https://doi.org/10.3390/molecules24234239.
35. Wang ZY, Wang P, Hillard CJ, Bjorling DE. Attenuation of cystitis and pain sensation in mice lacking fatty acid amide hydrolase. J Mol Neurosci. 2015;55(4):968–76. https://doi.org/10.1007/s12031-014-0453-x.

Musculoskeletal and Connective Tissues 8

8.1 ECS Distribution in the Bone

Despite bone's solid appearance, it is in actuality a dynamic tissue with constant remodeling. Osteoclasts secrete acid and protease to break down existing bones, while osteoblasts secrete collagen type 1 to form the structure of bones along with other proteins to promote mineral deposition. Through this process, our bones remain strong and respond constantly to the stresses and demands we place on them. Bone cells, including mesenchymal stem cells, express CB1 and CB2 receptors as well as TRPV1 and are capable of producing AEA and 2-AG [1].

8.2 The ECS and Osteoporosis

Activation of CB1 receptor in bones stimulates osteoclasts to break bone down. Studies in mice without CB1 receptors initially had higher peak bone density as would be expected with increased osteoclast activity but then went on to develop osteoporosis, a disease characterized by low bone mineral density (BMD). CB1 receptor activation is necessary for ongoing osteoblast differentiation, and without it, marrow stromal cells were more likely to form fat rather than osteoblasts, a part of the pathophysiology of osteoporosis due to aging [2].

CB2 activation on osteoblasts stimulates osteogenesis, while TRPV1 activation decreased it. Osteoblasts also release a transcription factor, RANK-L, that drives osteoblast differentiation. RANK-L release was significantly reduced by both CB2 and TRPV1 activation [3]. Here, we see clear evidence of homeostasis with different components of the ECS balancing the effect of the others, despite sharing common ligands. Unsurprisingly, a review of many studies looking at the effect of cannabis on BMD showed no effect in some cases, potential detrimental effects in others [4]. However, single nucleotide polymorphisms in CB2 gene have been linked to the development of osteoporosis. One study showed that a number of

J.-P. H. Dedam et al., *Diseases of the Endocannabinoid System*,
https://doi.org/10.1007/978-3-031-87484-0_8

missense variants resulted in higher risk of developing osteoporosis and would be expected to result in decreased function of CB2 receptor. Similar SNPs for CB1 have not been found [5]. Another study in Korean women with osteoporosis also linked seven SNPs in CB2 gene to lower BMD [6]. An additional 39 SNPs were also identified among Han Chinese women with the same effect [7]. Given this evidence, some fraction of osteoporosis may be due to an endocannabinoid deficiency. Even though THC is likely not a good osteoporosis drug, as noted above, a selective CB2 agonist could represent a very good therapeutic option, with the added benefit of its anti-inflammatory and antinociceptive effects. Another well-established cause of osteoporosis is the chronic glucocorticoid administration, in part through increased RANK-L release. A study using osteoclasts in vitro from healthy women showed that methylprednisolone caused dysregulation of the ECS, with inhibition of CB2 activity and increased TRPV1 activity. This again raises the possibility that CB2 agonists could also reverse osteoporosis from glucocorticoid exposure [8].

8.3 The ECS and Cartilage, Tendons, Fascia, and Synovium

Moving on to a less rigid subject, the cartilage, tendons, fascia, and synovium are all made of cartilage and maintained by fibroblasts. These cells express both CB1 and CB2 receptors and are capable of producing and degrading endocannabinoids. CB1 receptor is present in muscle tissues, although mostly at the neuromuscular junction [9]. CB1 activation here by 2-AG inhibits release activated by the muscarinic acetylcholine receptors [10]. This accounts for cannabis's effect as a muscle relaxer and has shown consistent benefit in the treatment of spasticity in muscle sclerosis [11]. It is possible that people with low ECS function may be more prone to muscle spasm. Activation of CB2 receptor on human gingival fibroblasts resulted in decreased release of inflammatory factors such as TNF, IL-2, and interferon-gamma, which may be of use in the treatment of gingivitis [12]. Neuron growth and proliferation mediated by fibroblast growth factor is also facilitated by CB1 receptor, which raises the possibility that the ECS may be involved in fibroblast proliferation as well [13].

8.4 ECS and Osteoarthritis

The best studied effects of the ECS are on the synovium and cartilage in osteoarthritis and rheumatoid arthritis. Experiments have shown that activation of cannabinoid receptors in chondrocytes by a synthetic agonist of both receptors caused a decrease in production of interleukin 1 beta, a signal which upregulates matrix metalloproteinases, leading to cartilage degradation [14]. In subsequent studies, cannabinoid receptors were also found to decrease IL-1alpha release from chondrocytes, resulting in decreased cartilage degradation [15]. Loss of cartilage is the main pathologic process in osteoarthritis; this breakdown can be precipitated by trauma, inflammation, and aging, among other factors. The balance between cartilage production and degradation becomes imbalanced, leading to more destruction and loss of chondrocytes. The subchondral bone also begins to remodel with formation of vascular channels that allow more inflammatory cells to reach the joint, leading to more destruction of cartilage [16]. The ECS is beneficially involved in every step of the pathogenesis of degenerative arthritis, slow degradation of cartilage, decreasing inflammation through activation of CB2 receptor on immune cells, and decreasing bone reabsorption via the same CB2 receptor activation. Early suppression of inflammation by a FAAH inhibitor was shown in mice models to decrease end-stage OA as well as prevent the development of chronic pain. The role of the ECS in modulating pain will be covered separately in the chapter on the peripheral nervous system [17]. Furthermore, CB2 knockout mice have been shown to develop far more severe OA than wild-type mice, again indicating that some patients with early onset of severe OA may actually suffer from an underperforming ECS, specifically CB2 activity.

8.5 Rheumatoid Arthritis

The synovium of the joint can also play a major role in inflammation after injury, producing TNF and IL-1 alpha and beta that can further drive loss of cartilage as seen above. In a study on mice that had their ACLs traumatically ruptured, CB2 receptors were unregulated in the synovium. Activation of those receptors led to the decreased release of TNF-1 and IL-1 alpha and beta as well as attenuation of the immune system [18]. A separate study showed not only a similar decrease in IL-6 and IL-8 production from synovial cells but that TPRV1 was also involved in decreasing inflammation [19]. Rheumatoid arthritis (RA) is an autoimmune disease characterized by immunological attack and severe inflammation of the synovium, leading to joint degradation. CB2 is unregulated in RA and 2-AG and AEA are detectable in joint fluid of RA sufferers but not healthy joints [20]. The purpose of the upregulation of CB2 in RA, again, is feedback inhibition of inflammation signals with CB2 suppressing the same inflammatory cytokines that drove its own expression [21]. Limited studies on humans, specifically randomized control trial using Sativex, a combination THC/CBD oral spray, helped reduce pain and disease activity [22]. Unfortunately, that small study has done little to make the use of cannabinoids for RA mainstream, despite its obvious potential, especially since activation of CB2 on its own would not have any psychoactive effects.

8.6 ECS and Exercise

There is now reasonable evidence for increases in endocannabinoids in humans during and after exercise. These results are more most substantial for AEA with a bump of 75% above baseline levels with acute exercise. Many have even theorized that the “runner’s high” previously attributed to the release of endogenous opiates may actually be due to AEA. Of course, as we will see in subsequent chapters, both peripherally and centrally, the release of endogenous opiates and cannabinoids is inextricably linked. The purpose of this increase in AEA is not understood, but may be related to metabolic refueling, given AEA effect on glucose, fat stores, and appetite. It may also exist as a natural anesthetic following strenuous exertion [23].

References

1. Rossi F, Tortora C, Punzo F, Bellini G, Argenziano M, Di Paola A, Torella M, Perrotta S. The endocannabinoid/endovanilloid system in bone: from osteoporosis to osteosarcoma. Int J Mol Sci. 2019;20(8):1919. https://doi.org/10.3390/ijms20081919.
2. Idris AI, Sophocleous A, Landao-Bassonga E, Canals M, Milligan G, Baker D, et al. Cannabinoid receptor type 1 protects against age-related osteoporosis by regulating osteoblast and adipocyte differentiation in marrow stromal cells. Cell Metab. 2009;10(2):139–47. https://doi.org/10.1016/j.cmet.2009.07.006.
3. Rossi F, Bellini G, Tortora C, Bernardo ME, Luongo L, Conforti A, Starc N, Manzo I, Nobili B, Locatelli F, Maione S. CB(2) and TRPV(1) receptors oppositely modulate in vitro human osteoblast activity. Pharmacol Res. 2015;99:194–201. https://doi.org/10.1016/j.phrs.2015.06.010.
4. Magno LAV, Tameirão DR, Alves LF, Guimarães NS. Effect of recreational cannabis use on bone mineral density: a systematic review. Osteoporos Int. 2024;35(3):391–9. https://doi.org/10.1007/s00198-023-06992-4.
5. Karsak M, Cohen-Solal M, Freudenberg J, Ostertag A, Morieux C, Kornak U, Essig J, Erxlebe E, Bab I, Kubisch C, de Vernejoul MC, Zimmer A. Cannabinoid receptor type 2 gene is associated with human osteoporosis. Hum Mol Genet. 2005;14(22):3389–96. https://doi.org/10.1093/hmg/ddi370.
6. Woo JH, Kim H, Kim JH, Kim JG. Cannabinoid receptor gene polymorphisms and bone mineral density in Korean postmenopausal women. Menopause. 2015;22(5):512–9. https://doi.org/10.1097/GME.0000000000000339.
7. Zhang C, Ma J, Chen G, Fu D, Li L, Li M. Evaluation of common variants in CNR2 gene for bone mineral density and osteoporosis susceptibility in postmenopausal women of Han Chinese. Osteoporos Int. 2015;26(12):2803–10. https://doi.org/10.1007/s00198-015-3195-x.
8. Bellini G, Torella M, Manzo I, Tortora C, Luongo L, Punzo F, Colacurci N, Nobili B, Maione S, Rossi F. PKCβII-mediated cross-talk of TRPV1/CB2 modulates the glucocorticoid-induced osteoclast overactivity. Pharmacol Res. 2017;115:267–74. https://doi.org/10.1016/j.phrs.2016.11.039.
9. McPartland JM. Expression of the endocannabinoid system in fibroblasts and myofascial tissues. J Bodyw Mov Ther. 2008;12(2):169–82. https://doi.org/10.1016/j.jbmt.2008.01.004.
10. Newman Z, Malik P, Wu TY, Ochoa C, Watsa N, Lindgren C. Endocannabinoids mediate muscarine-induced synaptic depression at the vertebrate neuromuscular junction. Eur J Neurosci. 2007;25:1619–30.
11. Filippini G, Minozzi S, Borrelli F, Cinquini M, Dwan K. Cannabis and cannabinoids for symptomatic treatment for people with multiple sclerosis. Cochrane Database Syst Rev. 2022;5(5):CD013444. https://doi.org/10.1002/14651858.CD013444.pub2.
12. Abidi AH, Abhyankar V, Alghamdi SS, Tipton DA, Dabbous M. Phytocannabinoids regulate inflammation in IL-1β-stimulated human gingival fibroblasts. J Periodontal Res. 2022;57(6):1127–38. https://doi.org/10.1111/jre.13050.
13. Williams EJ, Walsh FS, Doherty P. The FGF receptor uses the endocannabinoid signaling system to couple to an axonal growth response. J Cell Biol. 2003;160(4):481–6. https://doi.org/10.1083/jcb.200210164.
14. Dunn SL, Wilkinson JM, Crawford A, Le Maitre CL, Bunning RA. Cannabinoid WIN-55,212-2 mesylate inhibits interleukin-1β induced matrix metalloproteinase and tissue inhibitor of matrix

metalloproteinase expression in human chondrocytes. Osteoarthr Cartil. 2014;22(1):133–44. https://doi.org/10.1016/j.joca.2013.10.016.

15. Mbvundula EC, Bunning RA, Rainsford KD. Arthritis and cannabinoids: HU-210 and Win-55,212-2 prevent IL-1alpha-induced matrix degradation in bovine articular chondrocytes in-vitro. J Pharm Pharmacol. 2006;58(3):351–8. https://doi.org/10.1211/jpp.58.3.0009.
16. Krustev E, Rioux D, McDougall JJ. Mechanisms and mediators that drive arthritis pain. Curr Osteoporos Rep. 2015;13(4):216–24. https://doi.org/10.1007/s11914-015-0275-y.
17. McDougall JJ, Muley MM, Philpott HT, Reid A, Krustev E. Early blockade of joint inflammation with a fatty acid amide hydrolase inhibitor decreases end-stage osteoarthritis pain and peripheral neuropathy in mice. Arthritis Res Ther. 2017;19(1):106. https://doi.org/10.1186/s13075-017-1313-1.
18. Rzeczycki P, Rasner C, Lammlin L, Junginger L, Goldman S, Bergman R, Redding S, Knights AJ, Elliott M, Maerz T. Cannabinoid receptor type 2 is upregulated in synovium following joint injury and mediates anti-inflammatory effects in synovial fibroblasts and macrophages. Osteoarthr Cartil. 2021;29(12):1720–31. https://doi.org/10.1016/j.joca.2021.09.003.
19. Lowin T, Pongratz G, Straub RH. The synthetic cannabinoid WIN55,212-2 mesylate decreases the production of inflammatory mediators in rheumatoid arthritis synovial fibroblasts by activating CB2, TRPV1, TRPA1 and yet unidentified receptor targets. J Inflamm (Lond). 2016;13:15. https://doi.org/10.1186/s12950-016-0114-7.
20. Fukuda S, Kohsaka H, Takayasu A, et al. Cannabinoid receptor 2 as a potential therapeutic target in rheumatoid arthritis. BMC Musculoskelet Disord. 2014;15:275.
21. Gui H, Liu X, Wang ZW, He DY, Su DF, Dai SM. Expression of cannabinoid receptor 2 and its inhibitory effects on synovial fibroblasts in rheumatoid arthritis. Rheumatology (Oxford). 2014;53(5):802–9. https://doi.org/10.1093/rheumatology/ket447.
22. Blake DR, Robson P, Ho M, Jubb RW, McCabe CS. Preliminary assessment of the efficacy, tolerability and safety of a cannabis-based medicine (Sativex) in the treatment of pain caused by rheumatoid arthritis. Rheumatology (Oxford). 2006;45(1):50–2. https://doi.org/10.1093/rheumatology/kei183.
23. Desai S, Borg B, Cuttler C, Crombie KM, Rabinak CA, Hill MN, Marusak HA. A systematic review and meta-analysis on the effects of exercise on the endocannabinoid system. Cannabis Cannabinoid Res. 2022;7(4):388–408. https://doi.org/10.1089/can.2021.0113.

The Immune System

9

The effect of the ECS on inflammation has been cited again and again throughout the earlier chapters of this book and is one of its most significant roles in the body. This effect is mediated through modulation of the immune system, both innate and adaptive. We will now go into further detail about the role of the ECS in immune function and its effect on all stages of the immune response.

9.1 ECS Distribution on Immune Cells

Nearly all immune cells express cannabinoid receptors, with B cells having the higher levels of expression, followed by natural killer cells, polymorphonuclear neutrophils, T8 cells, monocytes, and T4 cells [1]. CB2 is expressed at much higher levels in immune cells than CB1, and in the spleen and tonsils, the level of CB2 expression is as high as CB1 expression of the brain [2]. Immune cells are also capable of producing endocannabinoids, and in fact, a study in mice leukocytes showed that ANA could induce its own synthesis via CB2 receptor [3].

9.2 ECS and Immune Cell Differentiation

While the full role of the ECS on immune development has yet to elucidated, what follows is the current understanding. Immune cells originate in the bone marrow from hematopoietic stem cells. AEA has long been known to act as a hematopoietic stem cell growth factor via CB2 in cooperation with other known growth factors such as IL-3, erythropoietin, and granulocyte colony-stimulating factor [4]. Activation of CB2 receptor in mice following sublethal irradiation promoted hematopoietic stem cell recovery and enhanced the number them entering the cell cycle [5]. At the same time, in mouse MS models, CB2 activation decreased myeloid progenitor migration to the neural tissue, decreasing inflammation [6]. Immature B

J.-P. H. Dedam et al., *Diseases of the Endocannabinoid System*,
https://doi.org/10.1007/978-3-031-87484-0_9

cells are dependent on CB2 activation to remain in the bone marrow in order to complete maturation, and inhibition of this function decreased circulating mature B cells [7]. B cells have been shown to have variable levels of CB2 expression, with low levels seen during most of differentiation but then being dramatically upregulated in B-cell subsets after CD40 activation, indicating involvement of the ECS [8, 9]. CB2 expression levels are in turn regulated by lymphocytic production of TGF-beta, which decreases CB2 expression. TGF-beta is usually thought of an immunosuppressant. THC stimulates lymphocytic production of TGF-beta via CB2 activation, and this appears to be a mechanism by which CB2 regulates itself via a negative autocrine loop [10]. Exposure to THC in utero was observed to cause alterations in T-cell differentiation, although the mechanisms of this are unknown [11]. Given the ECS effects on cytokine production, which also drive immune cell differentiation, and important intracellular signaling pathways such as decreasing cAMP levels and activation of the MAPK pathway, it is no surprise that they are involved in the development of immune cells, but much more study is still needed to be done to understand its role more fully.

9.3 ECS and the Innate Immune System

We will now examine the various components of the immune system and further describe the role of the ECS, beginning with the innate immune system. This older portion of the immune system provides a nonspecific defense as opposed to the adaptive immune system, which uses pathogen-specific antibodies and MHC proteins to target their attack [12, 13].

9.3.1 Macrophages and Monocytes

Macrophages, called monocytes while still in the blood stream and often have another specific name depending on the tissue in which they are located, phagocytize and break down pathogens, dead cells, and other foreign debris. They also interact extensively with other components of the immune system and other surrounding cells via the exchange of cytokines. They are a major source of inflammatory signals such as nitric oxide, TNF, PGE2, and IL-6. THC has long been known to decrease the growth and phagocytotic activity of macrophages [14]. In a separate experiment using the same cell line, both 2-AG and AEA were capable of decreasing IL-6 production in response to bacterial LPS, and AEA was able to reduce NO production as well, while 2-AG actually slightly increased it. Anti-inflammatory effects were mediated by CB2 receptor in the case of both ligands, but it appears that 2-AG is more readily metabolized to arachidonic acid (AA) and PGE-2, both of which actually have pro-inflammatory effects and drive NO production despite CB2 activation [15]. Separately, 2-AG has also been shown to increase phagocytosis in human macrophages via CB2 [16]. This indicates that the control of inflammation, at least in macrophages, is not unidirectional and that the ECS is capable of providing homeostasis depending on which endocannabinoid is produced. The monocytes of chronic cannabis users have increased expression of both CB1 and CB2 receptors compared to nonusers, indicating the potential for immuno-modulation by chronic cannabinoid exposure prompting concern that it may be impairing immunity [17]. An 8-week study looking changes in immune function of various lymphocytes in HIV patients using medical cannabis did not find any statistical change in immune cell population or function in the short term [18]. Another study showed that heavy cannabis users among HAART-treated HIV patients actually benefited from reduced inflammation and reduced number frequency of monocytes without negative effects [19]. Another receptor involved in the endocannabinoid system, GPR55, is present in macrophages and is weakly activated by endocannabinoids and actually inhibited by CBD, which had a pro-inflammatory effect, with increased TNF and IL-12 production.

9.3.2 Natural Killer Cells

GPR55 is also found in natural killer cells, which are tasked with rapidly attacking cancer cells and cells infected with viruses. Activation of GPR55 here also boosted inflammation such as on macrophages increasing production of TNF and INF-gamma [12, 13]. 2-AG was also able induce natural killer cells to migrate, while THC and AEA did not [20]. CB2 knockout mice have been shown to have deficiencies of natural killer cells, indicating the importance of this receptor in their differentiation as well [21].

9.3.3 Neutrophils

Neutrophils are the most active cells in acute inflammation, drawn to sites of tissue damage or infection. This migration is inhibited by AEA via CB2 [22]. Migration of immune cells into tissues is vital for immune response, but it can also become a significant pathologic process as seen in many inflammatory diseases, and a better understanding what signals cause which cells to migrate and under what conditions could provide potential therapeutic targets [23]. CB2 activation was shown to decrease TNF and metalloprotienase-9 release from neutrophils in vascular wall plaques, which can drive atherosclerosis, although the receptor is not highly expressed there [24]. 2-AG was again shown to have the opposite effect, increasing inflammation, although in neutrophils, it is clearly related to hydrolysis of 2-AG into AA and independent of cannabinoid receptors [25].

9.3.4 Basophils and Eosinophils

The ECS has not been identified as of yet in basophils and little has been studied in eosinophils either. Eosinophil migration can be triggered by 2-AG via CB2 receptor, as opposed to inflammatory effects of 2-AG in neutrophils being mediated by pro-inflammatory metabolic products of 2-AG like AA and independent of CB2 receptor [26]. This same lab went on to show in mice that 2-AG levels increased in the skin after induction of contact dermatitis and that blocking CB2 receptor could attenuate dermatitis [27]. Subsequent work on human eosinophil migration in humans has shown that blockage of 2-AG breakdown to AA partially inhibited migration and that IL-5, a key activator of eosinophils, enhanced the effect of 2-AG [28]. While eosinophils are often implicated on immune disease, their main evolutionary function was to fight helminthic infections. Indeed, increases in both 2-AG and AEA were noted in rodents' intestinal and lung tissues infected with a nematode similar to hookworm but not in the blood stream. Increases in 2-AG levels were associated with decreased weight loss and decreased parasitic egg burden, indicating an important protective effect. Contrary to previous findings above, however, they found that blockage of CB1 receptor and not CB2 receptor worsened parasitic infection. CB1 blockage decreased IL-5 production but not other cytokines, which was noted above as being an important activator of eosinophil migration along with 2-AG. The nematodes also express cannabinoid receptors, however, so it is possible that some of this is beneficial effect is mediated by the worm's own receptors and not the immune system or potentially cross talk between the host and parasite ECSs [29].

9.3.5 Mast Cells

Mast cells are another set of leukocytes implicated in allergic reactions residing in the tissues themselves and also expressing both CB1 and CB2 receptors. PEA has also shown to have a regulatory effect on mast cells, although whether this is via CB2 or another receptor was not investigated in a study of spinal cord injury where PEA administration reduced mast cell infiltration and activation, reducing inflammation [30]. A study using human hair follicles showed that tonic CB1 activation controlled activation and maturation of mast cells. Insufficient ECS tone may therefore contribute to the excessive inflammation and pruritus seen in allergic and other mast cell-driven skin diseases [31]. The same inhibitory effect was also demonstrated in mucosal tissues of the nose, another important site of allergic diseases [32].

9.3.6 Dendritic Cells

Dendritic cells are the final component of the innate immune system to be discussed, although their main function is to collect and present antigens to the T and B cells of the adaptive immune system. They express both CB1 and CB2 and demonstrate increased production of 2-AG but not AEA after stimulation by bacterial LPS [33]. AEA in the blood stream caused dendritic cells to release fewer inflammatory cytokines, inhibiting induction of T-cell TH1 and TH17 lineages via CB2 activation in patient with MS [34]. 2-AG again prompted migration of dendritic cells, while AEA in the bone marrow lead to apoptosis but only after activation of both CB1 and CB2 receptors via the NF-KappaB pathway [35, 36]. Deletion of genes for CB1 and CB2 in mice models resulted in a more robust T-cell response [37].

9.4 ECS and Adaptive Immune System

9.4.1 T Cells

T cells coordinate the adaptive immune response and are split into a number of subtype based on the receptors expressed. T cells develop in the thymus and exposure to different cytokines determines their subtype. CD8+ or killer cells are capable of killing virally infected or cancer cells. CD4+ helper T cells activate B cells or CD8+ cells to fight anything they recognize as foreign. In early studies, it was noted that AEA was capable of inducing apoptosis in T cells and B cells [38]. AEA was later found to inhibit TNF-alpha induced by the NF-KappaB pathway, which activates pro-inflammatory genes in T cells prompting production and release of cytokines and differentiation of the T cells, in a CB2 receptor-independent fashion [39]. This effect may be mediated by binding PPAR gamma as 2-AG can suppress IL-2 secretion from activated T cells via PPAR gamma [40]. AEA has also been found to have an antiproliferative effect on human T cells via CB2 receptor and had a significant effect on IL-17 production, the predominant cytokine of T-helper cell subtype 17 or TH17, which is pro-inflammatory and implicated in a number of inflammatory and autoimmune diseases [41]. A relatively common polymorphism of CB2 receptor 188–189 GG/GG results in two times greater proliferation of T cells due to loss of inhibitory function and may be a risk factor for the development of autoimmune disease [42]. In fact, other CB2 polymorphisms such as a CB2 Q63R have already been linked to immune thrombocytopenia (ITP) where T cells directly attack platelets [43] in two separate studies [44]. The same polymorphism has also been linked to multiple sclerosis [45] and immune-mediated disease in patients with chronic HCV infection [46]. These findings raise the very real possibility that some proportion of autoimmune disease is due to ECS deficiency. Studies have further shown that the immune system is capable of modulating the ECS as well, with T-helper cell 2 (TH2) cytokines IL-4 and IL-10 reducing FAAH activity, while T helper cell 1 (TH1) cytokines IL-12 and interferon-gamma decreased FAAH function [47]. The first effect is decreasing endocannabinoid tone by clearing more AEA and the second is increasing it by allowing its accumulation. Additionally, activation of CB2 receptor has been shown to shift the balance of TH1 to TH2 response by modulating the production of cytokines [48]. This creates a great deal of cross talk between T cell and the ECS, two opposing feedforward activation loops. TH1 cells are more associated with a pro-inflammatory, antimicrobial effect, while TH2 cells are associated with anti-inflammatory response but also eosinophil and IgE activation, which in excess can lead to allergy and atrophy [49]. Evidence of medical cannabis effect on T-cell function was demonstrated in a study of patients in Israel undergoing treatment with checkpoint inhibitors for cancer. These drugs allow increased activation of T cells to fight cancer. Patients using a moderate dose of medical cannabis while on this therapy had decrease in time to tumor progression and lower overall survival, indicating overall inhibition of T-cell activity [50].

9.4.2 B Cells

Finally, B-lymphocytes, which produce antibodies, express the highest amount of CB2 receptor of all immune cells. Cannabinoids are capable of inducing proliferation of activated B cells via CB2 receptor [51]. Immature B cells in the bone marrow need to migrate to the spleen to finish differentiation. That migration is triggered in part by 2-AG via CB2 receptor. Myeloid leukemia cells also express CB2 and can be stimulated to migrate via the same mechanism [52]. As was noted above, CB2 receptor levels also drop during differentiation of B cells, before returning to previous levels after final differentiation into memory B cells, indicating a yet to be fully understood role in that process. B-cell expression of CB2 receptor also increased dramatically with CD40-mediated antigen-presenting cell activation in human tonsils [8, 9]. A role of the ECS in B-cell differentiation is supported by the finding that CB2 knockout mice are deficient in a number of B-cell subsets including marginal zone and peritoneal B cells [21]. In the case of the marginal B cells, CB2 activation by 2-AG resulted in their retention to this area, and blockage of CB2 receptor results in diminished humeral immunity [53]. As seen above in the shift of TH1 to TH2 cells, CB2 activation in B cells could also prompt a class switch from IgM to IgE antibody production, again associated with changes in the immune response [54]. While one study showed that cannabis users do not have higher rates of allergic symptoms, they did not have significantly higher levels of antibodies to common allergens including mold, dust mites, and cat dander [55].

The role of the ECS, particularly CB2 receptor, is vital to the development and function of the immune system. As seen above, we have glimpses of its role; however, a great deal of research still remains to be done to better understand that role, how disruption of the normal ECS can lead to diseases such as autoimmunity or allergy, and how the ECS can be modulated in those disease states to provide potential treatments.

References

1. Bouaboula M, Rinaldi M, Carayon P, Carillon C, Delpech B, Shire D, Le Fur G, Casellas P. Cannabinoid-receptor expression in human leukocytes. Eur J Biochem. 1993;214(1):173–80. https://doi.org/10.1111/j.1432-1033.1993.tb17910.x.
2. Galiègue S, Mary S, Marchand J, Dussossoy D, Carrière D, Carayon P, Bouaboula M, Shire D, Le Fur G, Casellas P. Expression of central and peripheral cannabinoid receptors in human immune tissues and leukocyte subpopulations. Eur J Biochem. 1995;232(1):54–61. https://doi.org/10.1111/j.1432-1033.1995.tb20780.x.

3. Pestonjamasp VK, Burstein SH. Anandamide synthesis is induced by arachidonate mobilizing agonists in cells of the immune system. Biochim Biophys Acta. 1998;1394(2–3):249–60. https://doi.org/10.1016/s0005-2760(98)00110-6.
4. Valk P, Verbakel S, Vankan Y, Hol S, Mancham S, Ploemacher R, Mayen A, Löwenberg B, Delwel R. Anandamide, a natural ligand for the peripheral cannabinoid receptor is a novel synergistic growth factor for hematopoietic cells. Blood. 1997;90(4):1448–57.
5. Jiang S, Yigong F, Avraham HK. Regulation of hematopoietic stem cell trafficking and mobilization by the endocannabinoid system. Transfusion. 2011;51:65S–71S.
6. Palazuelos J, et al. The CB2 cannabinoid receptor controls myeloid progenitor trafficking: involvement in the pathogenesis of an animal model of multiple sclerosis. J Biol Chem. 2008;283(19):13320–9.
7. Pereira J, An J, Xu Y, et al. Cannabinoid receptor 2 mediates the retention of immature B cells in bone marrow sinusoids. Nat Immunol. 2009;10:403–11.
8. Carayon P, et al. Modulation and functional involvement of CB2 peripheral cannabinoid receptors during B-cell differentiation. Blood J Am Soc Hematol. 1998b;92(10):3605–15.
9. Carayon P, Marchand J, Dussossoy D, Derocq JM, Jbilo O, Bord A, Bouaboula M, Galiègue S, Mondière P, Pénarier G, Fur GL, Defrance T, Casellas P. Modulation and functional involvement of CB2 peripheral cannabinoid receptors during B-cell differentiation. Blood. 1998a;92(10):3605–15.
10. Gardner B, Zu LX, Sharma S, Liu Q, Makriyannis A, Tashkin DP, Dubinett SM. Autocrine and paracrine regulation of lymphocyte CB2 receptor expression by TGF-beta. Biochem Biophys Res Commun. 2002;290(1):91–6. https://doi.org/10.1006/bbrc.2001.6179.
11. Lombard C, et al. Perinatal exposure to Δ9-tetrahydrocannabinol triggers profound defects in T cell differentiation and function in fetal and postnatal stages of life, including decreased responsiveness to HIV antigens. J Pharmacol Exp Ther. 2011;339(2):607–17.
12. Chiurchiù V, Battistini L, Maccarrone M. Endocannabinoid signalling in innate and adaptive immunity. Immunology. 2015a;144(3):352–64. https://doi.org/10.1111/imm.12441.
13. Chiurchiù V, Lanuti M, De Bardi M, Battistini L, Maccarrone M. The differential characterization of GPR55 receptor in human peripheral blood reveals a distinctive expression in monocytes and NK cells and a proinflammatory role in these innate cells. Int Immunol. 2015b;27(3):153–60. https://doi.org/10.1093/intimm/dxu097.
14. Tang JL, Lancz G, Specter S, Bullock H. Marijuana and immunity: tetrahydrocannabinol-mediated inhibition of growth and phagocytic activity of the murine macrophage cell line, P388D1. Int J Immunopharmacol. 1992;14(2):253–62. https://doi.org/10.1016/0192-0561(92)90037-l.
15. Chang YH, Lee ST, Lin WW. Effects of cannabinoids on LPS-stimulated inflammatory mediator release from macrophages: involvement of eicosanoids. J Cell Biochem. 2001;81(4):715–23. https://doi.org/10.1002/jcb.1103.
16. Gokoh M, Kishimoto S, Oka S, Sugiura T. 2-Arachidonoylglycerol enhances the phagocytosis of opsonized zymosan by HL-60 cells differentiated into macrophage-like cells. Biol Pharm Bull. 2007;30(7):1199–205. https://doi.org/10.1248/bpb.30.1199.
17. Nong L, et al. Altered cannabinoid receptor mRNA expression in peripheral blood mononuclear cells from marijuana smokers. J Neuroimmunol. 2002;127(1–2):169–76.
18. Bredt BM, Higuera-Alhino D, Shade SB, Hebert SJ, McCune JM, Abrams DI. Short-term effects of cannabinoids on immune phenotype and function in HIV-1-infected patients. J Clin Pharmacol. 2002;42(S1):82S–9S. https://doi.org/10.1002/j.1552-4604.2002.tb06007.x.
19. Manuzak JA, Gott TM, Kirkwood JS, et al. Heavy cannabis use associated with reduction in activated and inflammatory immune cell frequencies in antiretroviral therapy-treated human immunodeficiency virus-infected individuals. Clin Infect Dis. 2018;66(12):1872–82. https://doi.org/10.1093/cid/cix1116.
20. Kishimoto S, Muramatsu M, Gokoh M, Oka S, Waku K, Sugiura T. Endogenous cannabinoid receptor ligand induces the migration of human natural killer cells. J Biochem. 2005;137(2):217–23. https://doi.org/10.1093/jb/mvi021.

21. Ziring D, Wei B, Velazquez P, Schrage M, Buckley NE, Braun J. Formation of B and T cell subsets require the cannabinoid receptor CB2. Immunogenetics. 2006;58(9):714–25. https://doi.org/10.1007/s00251-006-0138-x.
22. McHugh D, Tanner C, Mechoulam R, Pertwee RG, Ross RA. Inhibition of human neutrophil chemotaxis by endogenous cannabinoids and phytocannabinoids: evidence for a site distinct from CB1 and CB2. Mol Pharmacol. 2008;73(2):441–50. https://doi.org/10.1124/mol.107.041863.
23. Luster AD, Alon R, von Andrian UH. Immune cell migration in inflammation: present and future therapeutic targets. Nat Immunol. 2005;6(12):1182–90. https://doi.org/10.1038/ni1275.
24. Montecucco F, Di Marzo V, da Silva RF, Vuilleumier N, Capettini L, Lenglet S, Pagano S, Piscitelli F, Quintao S, Bertolotto M, Pelli G, Galan K, Pilet L, Kuzmanovic K, Burger F, Pane B, Spinella G, Braunersreuther V, Gayet-Ageron A, Pende A, Viviani GL, Palombo D, Dallegri F, Roux-Lombard P, Santos RA, Stergiopulos N, Steffens S, Mach F. The activation of the cannabinoid receptor type 2 reduces neutrophilic protease-mediated vulnerability in atherosclerotic plaques. Eur Heart J. 2012;33(7):846–56. https://doi.org/10.1093/eurheartj/ehr449.
25. Chouinard F, Lefebvre JS, Navarro P, Bouchard L, Ferland C, Lalancette-Hébert M, Marsolais D, Laviolette M, Flamand N. The endocannabinoid 2-arachidonoyl-glycerol activates human neutrophils: critical role of its hydrolysis and de novo leukotriene B4 biosynthesis. J Immunol. 2011;186(5):3188–96. https://doi.org/10.4049/jimmunol.1002853.
26. Oka S, Ikeda S, Kishimoto S, Gokoh M, Yanagimoto S, Waku K, Sugiura T. 2-arachidonoylglycerol, an endogenous cannabinoid receptor ligand, induces the migration of EoL-1 human eosinophilic leukemia cells and human peripheral blood eosinophils. J Leukoc Biol. 2004;76(5):1002–9. https://doi.org/10.1189/jlb.0404252.
27. Oka S, Wakui J, Ikeda S, Yanagimoto S, Kishimoto S, Gokoh M, Nasui M, Sugiura T. Involvement of the cannabinoid CB2 receptor and its endogenous ligand 2-arachidonoylglycerol in oxazolone-induced contact dermatitis in mice. J Immunol. 2006;177(12):8796–805. https://doi.org/10.4049/jimmunol.177.12.8796.
28. Larose MC, Turcotte C, Chouinard F, Ferland C, Martin C, Provost V, Laviolette M, Flamand N. Mechanisms of human eosinophil migration induced by the combination of IL-5 and the endocannabinoid 2-arachidonoyl-glycerol. J Allergy Clin Immunol. 2014;133(5):1480–2. https://doi.org/10.1016/j.jaci.2013.12.1081.
29. Batugedara HM, Argueta D, Jang JC, Lu D, Macchietto M, Kaur J, Ge S, Dillman AR, DiPatrizio NV, Nair MG. Host- and helminth-derived endocannabinoids that have effects on host immunity are generated during infection. Infect Immun. 2018;86(11):e00441–18. https://doi.org/10.1128/IAI.00441-18.
30. Esposito E, Paterniti I, Mazzon E, Genovese T, Di Paola R, Galuppo M, Cuzzocrea S. Effects of palmitoylethanolamide on release of mast cell peptidases and neurotrophic factors after spinal cord injury. Brain Behav Immun. 2011;25(6):1099–112. https://doi.org/10.1016/j.bbi.2011.02.006.
31. Sugawara K, Bíró T, Tsuruta D, Tóth BI, Kromminga A, Zákány N, Zimmer A, Funk W, Gibbs BF, Zimmer A, Paus R. Endocannabinoids limit excessive mast cell maturation and activation in human skin. J Allergy Clin Immunol. 2012;129(3):726–38. https://doi.org/10.1016/j.jaci.2011.11.009.
32. Sugawara K, Zákány N, Hundt T, Emelianov V, Tsuruta D, Schäfer C, Kloepper JE, Bíró T, Paus R. Cannabinoid receptor 1 controls human mucosal-type mast cell degranulation and maturation in situ. J Allergy Clin Immunol. 2013;132(1):182–93. https://doi.org/10.1016/j.jaci.2013.01.002.
33. Matias I, Pochard P, Orlando P, Salzet M, Pestel J, Di Marzo V. Presence and regulation of the endocannabinoid system in human dendritic cells. Eur J Biochem. 2002;269(15):3771–8. https://doi.org/10.1046/j.1432-1033.2002.03078.x.
34. Chiurchiù V, Cencioni MT, Bisicchia E, De Bardi M, Gasperini C, Borsellino G, Centonze D, Battistini L, Maccarrone M. Distinct modulation of human myeloid and plasmacytoid dendritic cells by anandamide in multiple sclerosis. Ann Neurol. 2013;May;73(5):626–36. https://doi.org/10.1002/ana.23875. Epub 2013 Apr 17. PMID: 23447381.

35. Maestroni GJ. The endogenous cannabinoid 2-arachidonoyl glycerol as in vivo chemoattractant for dendritic cells and adjuvant for Th1 response to a soluble protein. FASEB J. 2004;18(15):1914–6. https://doi.org/10.1096/fj.04-2190fje.
36. Do Y, McKallip RJ, Nagarkatti M, Nagarkatti PS. Activation through cannabinoid receptors 1 and 2 on dendritic cells triggers NF-kappaB-dependent apoptosis: novel role for endogenous and exogenous cannabinoids in immunoregulation. J Immunol. 2004;173(4):2373–82. https://doi.org/10.4049/jimmunol.173.4.2373.
37. Karmaus PW, Chen W, Crawford RB, Harkema JR, Kaplan BL, Kaminski NE. Deletion of cannabinoid receptors 1 and 2 exacerbates APC function to increase inflammation and cellular immunity during influenza infection. J Leukoc Biol. 2011;Nov;90(5):983–95. https://doi.org/10.1189/jlb.0511219. Epub 2011 Aug 26. PMID: 21873455; PMCID: PMC3206470.
38. Schwarz H, Blanco FJ, Lotz M. Anandamide, an endogenous cannabinoid receptor agonist inhibits lymphocyte proliferation and induces apoptosis. J Neuroimmunol. 1994;55(1):107–15. https://doi.org/10.1016/0165-5728(94)90152-x.
39. Sancho R, Calzado MA, Di Marzo V, Appendino G, Muñoz E. Anandamide inhibits nuclear factor-kappaB activation through a cannabinoid receptor-independent pathway. Mol Pharmacol. 2003;63(2):429–38. https://doi.org/10.1124/mol.63.2.429.
40. Rockwell CE, Snider NT, Thompson JT, Vanden Heuvel JP, Kaminski NE. Interleukin-2 suppression by 2-arachidonyl glycerol is mediated through peroxisome proliferator-activated receptor gamma independently of cannabinoid receptors 1 and 2. Mol Pharmacol. 2006;70(1):101–11. https://doi.org/10.1124/mol.105.019117.
41. Cencioni MT, Chiurchiù V, Catanzaro G, Borsellino G, Bernardi G, Battistini L, Maccarrone M. Anandamide suppresses proliferation and cytokine release from primary human T-lymphocytes mainly via CB2 receptors. PLoS One. 2010;5(1):e8688. https://doi.org/10.1371/journal.pone.0008688.
42. Sipe JC, Arbour N, Gerber A, Beutler E. Reduced endocannabinoid immune modulation by a common cannabinoid 2 (CB2) receptor gene polymorphism: possible risk for autoimmune disorders. J Leukoc Biol. 2005;78(1):231–8. https://doi.org/10.1189/jlb.0205111.
43. Mahmoud Gouda H, Mohamed Kamel NR. Cannabinoid CB2 receptor gene (CNR2) polymorphism is associated with chronic childhood immune thrombocytopenia in Egypt. Blood Coagul Fibrinolysis. 2013;24(3):247–51. https://doi.org/10.1097/MBC.0b013e32835aba1d.
44. Rossi F, Mancusi S, Bellini G, Roberti D, Punzo F, Vetrella S, Matarese SM, Nobili B, Maione S, Perrotta S. CNR2 functional variant (Q63R) influences childhood immune thrombocytopenic purpura. Haematologica. 2011;96(12):1883–5. https://doi.org/10.3324/haematol.2011.045732.
45. Tahamtan A, Rezaiy S, Samadizadeh S, Moradi A, Tabarraei A, Javid N, Oladnabi M, Naeimi MH. Cannabinoid CB2 receptor functional variation (Q63R) is associated with multiple sclerosis in Iranian subjects. J Mol Neurosci. 2020;70(1):26–31. https://doi.org/10.1007/s12031-019-01395-9.
46. Coppola N, Zampino R, Bellini G, Stanzione M, Capoluongo N, Marrone A, Macera M, Adinolfi LE, Giudice EM, Gentile I, Sagnelli E, Rossi F. CB2-63 polymorphism and immune-mediated diseases associated with HCV chronic infection. Dig Liver Dis. 2016;48(11):1364–9. https://doi.org/10.1016/j.dld.2016.07.005.
47. Maccarrone M, Valensise H, Bari M, Lazzarin N, Romanini C, Finazzi-Agrò A. Progesterone up-regulates anandamide hydrolase in human lymphocytes: role of cytokines and implications for fertility. J Immunol. 2001;166(12):7183–9. https://doi.org/10.4049/jimmunol.166.12.7183.
48. Yuan M, Kiertscher SM, Cheng Q, Zoumalan R, Tashkin DP, Roth MD. Delta 9-tetrahydrocannabinol regulates Th1/Th2 cytokine balance in activated human T cells. J Neuroimmunol. 2002;133(1–2):124–31. https://doi.org/10.1016/s0165-5728(02)00370-3.
49. Berger A. Th1 and Th2 responses: what are they? BMJ. 2000;321(7258):424. https://doi.org/10.1136/bmj.321.7258.424.

50. Bar-Sela G, Cohen I, Campisi-Pinto S, Lewitus GM, Oz-Ari L, Jehassi A, Peer A, Turgeman I, Vernicova O, Berman P, Wollner M, Moskovitz M, Meiri D. Cannabis consumption used by cancer patients during immunotherapy correlates with poor clinical outcome. Cancers (Basel). 2020;12(9):2447. https://doi.org/10.3390/cancers12092447.
51. Derocq JM, Ségui M, Marchand J, Le Fur G, Casellas P. Cannabinoids enhance human B-cell growth at low nanomolar concentrations. FEBS Lett. 1995;369(2–3):177–82. https://doi.org/10.1016/0014-5793(95)00746-v.
52. Jordà MA, Verbakel SE, Valk PJ, Vankan-Berkhoudt YV, Maccarrone M, Finazzi-Agrò A, Löwenberg B, Delwel R. Hematopoietic cells expressing the peripheral cannabinoid receptor migrate in response to the endocannabinoid 2-arachidonoylglycerol. Blood. 2002;99(8):2786–93. https://doi.org/10.1182/blood.v99.8.2786.
53. Muppidi JR, Arnon TI, Bronevetsky Y, Veerapen N, Tanaka M, Besra GS, Cyster JG. Cannabinoid receptor 2 positions and retains marginal zone B cells within the splenic marginal zone. J Exp Med. 2011;208(10):1941–8. https://doi.org/10.1084/jem.20111083.
54. Agudelo M, Newton C, Widen R, Sherwood T, Nong L, Friedman H, Klein TW. Cannabinoid receptor 2 (CB2) mediates immunoglobulin class switching from IgM to IgE in cultures of murine-purified B lymphocytes. J Neuroimmune Pharmacol. 2008;3(1):35–42. https://doi.org/10.1007/s11481-007-9088-9.
55. Min JY, Min KB. Marijuana use is associated with hypersensitivity to multiple allergens in US adults. Drug Alcohol Depend. 2018;182:74–7. https://doi.org/10.1016/j.drugalcdep.2017.09.039.

Peripheral Nervous System and Spinal Cord 10

We can now turn our attention to the nervous system, probably the best studied area of influence for the endocannabinoid system. We will begin with the peripheral nervous system, including the spinal cord, before moving to the rest of the central nervous system, namely, the brain.

10.1 Distribution of ECS in PNS

CB1 receptors are not as prevalent in the PNS as in the CNS; however, they are highly concentrated at nerve endings. These receptors have long been known to have antinociceptive effects in mice and rats, based on many unfortunate mice and rat torture experiments. Based on these early experiments, it became clear that the ECS had a greater effect on reducing nociception in chronic pain versus acute pain models [1].

CB2 receptors, generally associated with the immune system and not expressed extensively in the brain, are found in the spinal cord and dorsal root ganglia [2], specifically on the terminals of the somatic sensory nerves in the dorsal horn of the spinal cord [3]. TRPV1 is even more extensively expressed on peripheral nerve tissues present on small to medium diameter sensory neurons in the skin and reacts to heat, acid, and irritants like capsaicin. It is as well found in unmyelinated slow conducting fibers (C-fibers) and in the dorsal root ganglion [4]. It is also found in the spinal cord, specifically in the dorsal horn [5]. PPAR receptors are expressed throughout the nervous system and have been found to be active in a number of demyelinating and neuropathic conditions [6]. Nerve cells are also fully capable of producing and metabolizing both AEA and 2-AG, along with PEA [7].

J.-P. H. Dedam et al., *Diseases of the Endocannabinoid System*,
https://doi.org/10.1007/978-3-031-87484-0_10

10.2 Peripheral Nociception and Chronic Pain

The main focus of this chapter will be on the role of the ECS in pain at the level of peripheral nerves and the spinal cord. Nociception is the detection of tissue damage. Acute pain begins with stimulation and depolarization of nociceptors on sensory nerves. Lightly myelinated A-gamma fibers rapidly transmit pinpoint, sharp pain while unmyelinated C fibers transit more diffuse signals of burning and dull pain sensations. Large myelinated A-alpha and beta fibers transmit vibration and touch but, under pathological conditions, can carry aberrant pain signals. The cell bodies of peripheral neurons form the dorsal root bodies. Primary sensory neurons then synapse in the dorsal horn of the spinal cord with secondary neurons that carry the signal up the spinal cord to the brain. Neuromodulatory neurons also descend from the CNS here to affect activity at the synapse of the primary and secondary neurons [8]. The development and pathophysiology of chronic pain are more complex and can occur via a wide variety of different mechanisms occurring at the level of the DRG, spinal cord, and higher but is not fully understood. Some mechanisms involve increased sensitivity to nociception, also known as hyperalgesia. Sensitization of nerves can occur because of injury, either mechanical or chemical or via inflammation. Injured cells prompt the release of nerve growth factor and kinins that can worsen sensitivity. Some injuries prevent nerves from functioning correctly afterward, resulting in aberrant signaling that can be painful. Activation of the autonomic nervous system can also cause pain sensitization as can descending centrally mediated pain sensitization [9]. The ECS plays a role in all of these mechanisms and, in most cases, functions to prevent nerve sensitization [10]. It is an important target for treatment of chronic pain, and its deficiency may contribute to why some patients develop chronic pain while others do not.

10.3 ECS's Role in Peripheral Nociception and Neuropathy

Peripheral administration of AEA in animal models, where chemical damage of foot pads was used to elicit pain (torture), reduced patient's pain behaviors via CB1 receptor activation on peripheral neurons. This study also found that PEA had a similar effect via CB2 receptors and that co-administration had a substantial synergistic effect [11]. Local injection of 2-AG into foot pads of rats after previous injection with formalin (torture) also decreased nociception but this time, only via CB2 receptor. This effect was also seen, not surprisingly, with the injection of a MAGL inhibitor [12]. MAGL blockade was also capable of blocking capsaicin-induced pain [13] and inflammatory pain [14]. In some circumstances, this action is mediated peripherally by CB1 and CB2 receptor's activated release of norepinephrine (NE), as blocking adrenoreceptors inhibited the antinociceptive effects of AEA and PEA in a rat paw pressure model (torture) [15]. This is interesting in light of the fact that NE can have both pro-nociceptive or antinociceptive effects. The antinociceptive effects are via the release of endogenous opiates and possibly via an anti-inflammatory effect on immune cells [16]. In the previous chapter, we saw that CB2 was upregulated in response to inflammation, and sciatic nerve ligation models (torture) have demonstrated the same finding on injured nerves with buildup of the receptor at the spinal cord synapses but also along the axon itself [3]. And since increased ECS activation and signaling, as seen above, decreased pain transmission, upregulation of CB2 receptors after injury appears to be a physiologic response to decrease pain transmission and prevent nerve sensitization to pain. Another potential peripheral effect of the ECS on pain may be through stimulation of keratinocytes to produce endogenous opiates via activation of the CB2 receptor [17].

10.4 ECS's Role in Nociception and Neuropathy at the Level of the Dorsal Root Ganglia

Cannabis has recently shown potential at decreasing symptoms and potentially even preventing the development of oxaliplatin-induced peripheral neuropathy. In the dorsal root ganglia (DRG), while CB1 and CB2 are not extensively expressed there, CB1 receptor expression was increased in animal models treated with oxaliplatin, and its activation reduced subsequent mechanical hyperalgesia and cold allodynia [18]. Both CB1 and CB2 activation in rats with neuropathy prevented calcium influx in the DRG in response to the administration of capsaicin, and CB2 was found to have a similar effect at the level of the spinal cord in this experiment as well [19]. Experiments looking at neuropathy induced by cisplatin in rats also showed that ECS tone was increased at the level of spinal cord, with increases in AEA and 2-AG concentrations. Blocking FAAH in these experiments was as effective as morphine in reducing pain and superior to both gabapentin and amitriptyline. CB1 and CB2 receptor blockage impaired the effect, but also unexpectedly blockage of TRPV1 receptor had the same effect but only in the case of FAAH, not MAGL, blockage. This would indicate that there is an interaction between AEA and TRPV1 receptor at the level of the spinal cord that reduces neuropathic pain [20]. Human polymorphism in FAAH gene have been associated with increased sensitivity to pain, with the patients carrying both copies of the FAAH allele rs324420, which is known to have reduced activity, having significantly lower cold sensitivity and requiring less postoperative anesthesia. Patients with FAAH SNP rs4141964 AA also had reduced cold sensitivity [21]. In a separate study, patients with SNP rs4141964, TT had increased cold sensitivity at baseline [22]. In another human study looking at chronic back pain, CB2 was upregulated initially. In patients who went on to develop chronic back, however, only TRPV1 and FAAH remained elevated, with SNP in FAAH gene associated with mechanical and cold sensitivities: rs324420, rs4141964, and rs932816 specifically [23]. TPV1 receptors present in the DRG as well as the spinal cord are known to function in the transmission of various pain signals to the CNS and are involved in pathophysiology of chemotherapy-induced neuropathy and are unregulated in diabetic neuropathy models and after spinal cord and peripheral nerve injury and also in hyperalgesia after chronic opiate use. The exact mechanism of how AEA converts TRPV1 from a pro-neuropathy receptor to serving to prevent it is unknown but may involve desensitization of the receptor at higher concentrations of AEA and is discussed below [24]. As seen in the above experiments, boosting AEA by limiting FAAH activity appears to play an important role in the prevention of chronic pain and appears to involve TPRV1. PEA and OEA are also capable of binding TRPV1 and may play a role here as well [25]. Lack of necessary ECS tone may contribute in the development of all of the above neuropathic conditions as it plays a vital role in preventing them from developing in the first place. Patients with an underperforming ECS, as seen in cases of FAAH SNPs, would therefore be at greater risk and more vulnerable after trigger to develop chronic pain and painful neuropathies. In the case of idiopathic neuropathies, some of these may also be undiagnosed ECS deficiency syndromes. Supplementation with phyto-cannabinoids is effective in treating all of the above conditions and, in the case of chemotherapy-induced neuropathy, possibly even preventing it.

10.5 ECS's Role in Nociception and Neuropathy at the Level of the Spinal Cord

At the level of the spinal cord, administration of AEA has an expected pain-relieving effect. In rats, in low doses, the pain-relieving effect could be prevented by CB1 receptor blockage; however, at higher doses, the pain-relieving effect was prevented only by TRPV1 blockage [26]. AEA does have a lower binding affinity to CB1 receptor than the TRPV1, but that is expected. The experiment does bring up the possibility that only with sufficiently high ECS tone does TRPV1 switch from serving to facilitate transmission of pain to blocking it. In rat models of surgical pain, spinal levels of AEA are lower initially and increase as pain decreases. Both CB1 and CB2 were necessary to prevent excessive activation of astrocyte-driven inflammation [27]. Separate studies using ligation of the sciatic nerve of rats also showed increases in 2-AG in the spinal cord [28] and that these increases prevented pain transmission [29]. Similar injuries also upregulate CB1 receptors in the spinal cord [30] as well as CB2 receptors [31]. Vitamin D-deficient mice with allodynia have been shown to have reduced expression of CB1-CB2 in their spinal cords and that administration of PEA reversed allodynia [32]. The underlaying mechanism of these effects is not fully understood. Hemopressin, a pepcan, also shows benefit in preventing neuropathy. Oral administration to rats prevented hyper-analgesia with apparent effect in the dorsal horn of the spinal cord and DRG neurons independent of descending inhibitory pathways and its effect boosted by increased production of AEA [33]. CB2 knockout mice had increased activation of microglia and astrocytes, leading to the development of neuropathic pain after sciatic nerve injury, indicating that reductions in inflammation at the level of the spinal cord had an antinociceptive effect [34]. Intrathecal administration of CB1 receptor blockers results in the development of hyperalgesia in the absence of injury, indicating a tonic role of the ECS in the prevention of pain sensitization. Blocking NMDA receptors partially prevented sensitization, suggesting that that part of the role of the ECS in the spinal cord is inhibition of NMDA-dependent pain signaling as well [35]. Intrathecal administration of AEA also triggered the release of dynorphin A, an endogenous opiate, and the antinociceptive response could be blocked by both CB1 receptor antagonists and kappa opiate receptor antagonist [36]. Older experiments showed that that part of the antinociceptive effective in the spinal cords also mediated the activation of descending neurons from the CNS as evidenced by pain relief from injection of CB1 agonist in the ventricle and partial blockage of effect in rats with their spinal cords severed above the area of study [37]. Again, we see the importance of the ECS in limiting pain after injury but also in preventing neuropathic pain conditions at the level of the spinal cord and providing further evidence that ECS deficiency could contribute to the development of a variety of different neuropathic pain conditions. And, given that the tonic activity of the ECS is necessary for the prevention of hyperalgesia, ECS deficiency may also affect differences seen in patient's pain thresholds. This is enforced again by previously discussed finding that patients with chronic lower back pain have higher expression of TRPV1 and FAAH than patients without chronic pain [23] Also, irritable bowel syndrome, a condition characterized by increased visceral hypersensitivity, is also associated with a CB1 receptor polymorphism resulting in lower function [38].

10.6 Synergy with Opiate Receptors

As mentioned above, the ECS was capable of both boosting the release of endogenous opiates and preventing hyper-analgesia caused by chronic exogenous opiate administration. Beyond this, animal models have long demonstrated analgesic synergy of co-administered cannabinoids and opiates. Low-dose combinations of the two when given to mice in an acute pain model produced a greater level of analgesia than simple additivity would predict [39]. In a related acute pain model in rats, induction of a diabetic state decreased the pain-relieving effect of morphine due to a decreased release of endogenous opiates. THC administration, in the same rats, did not see a decrease in the release of endogenous opiates and showed preserved analgesia [40]. These findings, along with previously demonstrated ability to decrease opiate tolerance, make cannabinoids a potentially important opiate sparing therapy, allowing for similar pain relief with reduced doses of opiates. Both opiate and cannabinoid receptors are present in the same cells, so how does this interaction occur? Studies in rats have demonstrated that CB1 activation, and to a lesser degree CB2 activation, is necessary for the peripheral antinociceptive effect of morphine. Blocking CB1 prevented its effect, and blocking FAAH and therefore boosting the ECS increased the effect of morphine. Both receptors are also G-protein-coupled receptors that act by decreasing cAMP intracellular concentrations, so when present on the same cell as they often are, there may also be an additive effect within the intracellular signaling mechanism [41]. A study in rats demonstrated that administration of THC increased the density of mu-opioid receptors centrally, which could account both for synergy and the decrease in opiate tolerance with administration of cannabinoids [42]. Exogenous cannabinoids such as THC have been shown in studies to induce the release of the endogenous opiate dynorphin B in the spinal cord, although AEA did not, likely because it is degraded too quickly [43]. Chronic administration of THC leads to upregulation of genes to produce dynorphin and enkephalin, another endogenous opiate, in the spinal cords of mice [44]. Older experiments have hinted at potential allosteric binding of cannabinoids to opiate receptors, as well as heterodimer formations between both receptors on the cell membrane, although both of these interactions lead to decreased signaling [45]. While there is always concern in the mixing of psychoactive medications, especially when both are potential drugs of abuse, many patients report the opiate sparing benefits of cannabis, report using cannabinoids to wean off opiates, or even prefer cannabinoids to opiates for pain relief. Self-reported cannabis use in two small studies in postoperative patients did not result in significant change in outcomes, either positive or negative, so further study is still warranted [46, 47].

References

1. Pertwee RG. Cannabinoid receptors and pain. Prog Neurobiol. 2001;63(5):569–611. https://doi.org/10.1016/s0301-0082(00)00031-9.
2. Beltramo M, Bernardini N, Bertorelli R, Campanella M, Nicolussi E, Fredduzzi S, Reggiani A. CB2 receptor-mediated antihyperalgesia: possible direct involvement of neural mechanisms. Eur J Neurosci. 2006;23(6):1530–8. https://doi.org/10.1111/j.1460-9568.2006.04684.x.
3. Wotherspoon G, Fox A, McIntyre P, Colley S, Bevan S, Winter J. Peripheral nerve injury induces cannabinoid receptor 2 protein expression in rat sensory neurons. Neuroscience. 2005;135(1):235–45. https://doi.org/10.1016/j.neuroscience.2005.06.009.
4. Sun ZC, Ma SB, Chu WG, Jia D, Luo C. Canonical transient receptor potential (TRPC) channels in nociception and pathological pain. Neural Plast. 2020;2020:3764193. https://doi.org/10.1155/2020/3764193.
5. Zhou HY, Chen SR, Chen H, Pan HL. The glutamatergic nature of TRPV1-expressing neurons in the spinal dorsal horn. J Neurochem. 2009;108(1):305–18. https://doi.org/10.1111/j.1471-4159.2008.05772.x.
6. Villapol S. Roles of peroxisome proliferator-activated receptor gamma on brain and peripheral inflammation. Cell Mol Neurobiol. 2018;38(1):121–32. https://doi.org/10.1007/s10571-017-0554-5.
7. Woodhams SG, Sagar DR, Burston JJ, Chapman V. The role of the endocannabinoid system in pain. Handb Exp Pharmacol. 2015;227:119–43. https://doi.org/10.1007/978-3-662-46450-2_7.
8. Babos MB, Grady B, Wisnoff W, McGhee C. Pathophysiology of pain. Dis Mon. 2013;59(10):330–58. https://doi.org/10.1016/j.disamonth.2013.05.003.
9. Chapman CR, Vierck CJ. The transition of acute postoperative pain to chronic pain: an integrative overview of research on mechanisms. J Pain. 2017;18(4):359. https://doi.org/10.1016/j.jpain.2016.11.004.
10. Burston JJ, Woodhams SG. Endocannabinoid system and pain: an introduction. Proc Nutr Soc. 2014;73(1):106–17. https://doi.org/10.1017/S0029665113003650.
11. Calignano A, La Rana G, Giuffrida A, Piomelli D. Control of pain initiation by endogenous cannabinoids. Nature. 1998;394(6690):277–81. https://doi.org/10.1038/28393.
12. Guindon J, Desroches J, Beaulieu P. The antinociceptive effects of intraplantar injections of 2-arachidonoyl glycerol are mediated by cannabinoid CB2 receptors. Br J Pharmacol. 2007;150(6):693–701. https://doi.org/10.1038/sj.bjp.0706990.

13. Spradley JM, Guindon J, Hohmann AG. Inhibitors of monoacylglycerol lipase, fatty-acid amide hydrolase and endocannabinoid transport differentially suppress capsaicin-induced behavioral sensitization through peripheral endocannabinoid mechanisms. Pharmacol Res. 2010;62:249–58.
14. Guindon J, Guijarro A, Piomelli D, et al. Peripheral antinociceptive effects of inhibitors of monoacylglycerol lipase in a rat model of inflammatory pain. Br J Pharmacol. 2011;163:1464–78.
15. Romero TR, Resende LC, Guzzo LS, Duarte ID. CB1 and CB2 cannabinoid receptor agonists induce peripheral antinociception by activation of the endogenous noradrenergic system. Anesth Analg. 2013;Feb;116(2):463–72. https://doi.org/10.1213/ANE.0b013e3182707859. Epub 2013 Jan 9. PMID: 23302980.
16. Pertovaara A. The noradrenergic pain regulation system: a potential target for pain therapy. Eur J Pharmacol. 2013;716(1–3):2–7. https://doi.org/10.1016/j.ejphar.2013.01.067.
17. Ibrahim MM, Porreca F, Lai J, Albrecht PJ, Rice FL, Khodorova A, Davar G, Makriyannis A, Vanderah TW, Mata HP, Malan TP Jr. CB2 cannabinoid receptor activation produces antinociception by stimulating peripheral release of endogenous opioids. Proc Natl Acad Sci USA. 2005;102(8):3093–8. https://doi.org/10.1073/pnas.0409888102.
18. Pereira AF, Lisboa MRP, de Freitas Alves BW, da Silva CMP, Dias DBS, de Menezes KLS, Cesário FRAS, de França JC, de Oliveira AR, Hallak JEC, Zuardi AW, Crippa JA, de Alencar NMN, Lima-Júnior RCP, Vale ML. Endocannabinoid system attenuates oxaliplatin-induced peripheral sensory neuropathy through the activation of CB1 receptors. Neurotox Res. 2021;39:1782.
19. Sagar DR, Kelly S, Millns PJ, O'Shaughnessey CT, Kendall DA, Chapman V. Inhibitory effects of CB1 and CB2 receptor agonists on responses of DRG neurons and dorsal horn neurons in neuropathic rats. Eur J Neurosci. 2005;22(2):371–9. https://doi.org/10.1111/j.1460-9568.2005.04206.x.
20. Guindon J, Lai Y, Takacs SM, Bradshaw HB, Hohmann AG. Alterations in endocannabinoid tone following chemotherapy-induced peripheral neuropathy: effects of endocannabinoid deactivation inhibitors targeting fatty-acid amide hydrolase and monoacylglycerol lipase in comparison to reference analgesics following cisplatin treatment. Pharmacol Res. 2013;67(1):94–109. https://doi.org/10.1016/j.phrs.2012.10.013.
21. Cajanus K, Holmström EJ, Wessman M, Anttila V, Kaunisto MA, Kalso E. Effect of endocannabinoid degradation on pain: role of FAAH polymorphisms in experimental and postoperative pain in women treated for breast cancer. Pain. 2016;157(2):361–9. https://doi.org/10.1097/j.pain.0000000000000398.
22. Kim H, et al. Genetic predictors for acute experimental cold and heat pain sensitivity in humans. J Med Genet. 2006;43(8):e40.
23. Ramesh D, D'Agata A, Starkweather AR, Young EE. Contribution of endocannabinoid gene expression and genotype on low back pain susceptibility and chronicity. Clin J Pain. 2018;34(1):8–14. https://doi.org/10.1097/AJP.0000000000000508.
24. Spicarova D, Nerandzic V, Palecek J. Update on the role of spinal cord TRPV1 receptors in pain modulation. Physiol Res. 2014;63(Suppl 1):S225–36. https://doi.org/10.33549/physiolres.932713.
25. Benítez-Angeles M, Morales-Lázaro SL, Juárez-González E, Rosenbaum T. TRPV1: structure, endogenous agonists, and mechanisms. Int J Mol Sci. 2020;21(10):3421. https://doi.org/10.3390/ijms21103421.
26. Starowicz K, Makuch W, Osikowicz M, Piscitelli F, Petrosino S, Di Marzo V, Przewlocka B. Spinal anandamide produces analgesia in neuropathic rats: possible CB(1)- and TRPV1-mediated mechanisms. Neuropharmacology. 2012;62(4):1746–55. https://doi.org/10.1016/j.neuropharm.2011.11.021.
27. Alkaitis MS, Solorzano C, Landry RP, Piomelli D, DeLeo JA, Romero-Sandoval EA. Evidence for a role of endocannabinoids, astrocytes and p38 phosphorylation in the resolution of postoperative pain. PLoS One. 2010;5(5):e10891. https://doi.org/10.1371/journal.pone.0010891.

28. Petrosino S, Palazzo E, de Novellis V, Bisogno T, Rossi F, Maione S, Di Marzo V. Changes in spinal and supraspinal endocannabinoid levels in neuropathic rats. Neuropharmacology. 2007;52(2):415–22. https://doi.org/10.1016/j.neuropharm.2006.08.011.
29. Woodhams SG, Wong A, Barrett DA, Bennett AJ, Chapman V, Alexander SP. Spinal administration of the monoacylglycerol lipase inhibitor JZL184 produces robust inhibitory effects on nociceptive processing and the development of central sensitization in the rat. Br J Pharmacol. 2012;167(8):1609–19. https://doi.org/10.1111/j.1476-5381.2012.02179.x.
30. Lim G, Sung B, Ji RR, Mao J. Upregulation of spinal cannabinoid-1-receptors following nerve injury enhances the effects of Win 55,212-2 on neuropathic pain behaviors in rats. Pain. 2003;105(1–2):275–83. https://doi.org/10.1016/s0304-3959(03)00242-2.
31. Zhang J, Hoffert C, Vu HK, et al. Induction of CB2 receptor expression in the rat spinal cord of neuropathic but not inflammatory chronic pain models. Eur J Neurosci. 2003;17:2750–4.
32. Guida F, Boccella S, Belardo C, Iannotta M, Piscitelli F, De Filippis F, Paino S, Ricciardi F, Siniscalco D, Marabese I, Luongo L, Ercolini D, Di Marzo V, Maione S. Altered gut microbiota and endocannabinoid system tone in vitamin D deficiency-mediated chronic pain. Brain Behav Immun. 2020;85:128–41. https://doi.org/10.1016/j.bbi.2019.04.006.
33. Toniolo EF, Maique ET, Ferreira WA Jr, Heimann AS, Ferro ES, Ramos-Ortolaza DL, Miller L, Devi LA, Dale CS. Hemopressin, an inverse agonist of cannabinoid receptors, inhibits neuropathic pain in rats. Peptides. 2014;56:125–31. https://doi.org/10.1016/j.peptides.2014.03.016.
34. Racz I, Nadal X, Alferink J, Baños JE, Rehnelt J, Martín M, Pintado B, Gutierrez-Adan A, Sanguino E, Manzanares J, Zimmer A, Maldonado R. Crucial role of CB(2) cannabinoid receptor in the regulation of central immune responses during neuropathic pain. J Neurosci. 2008;28(46):12125–35. https://doi.org/10.1523/JNEUROSCI.3400-08.2008.
35. Richardson JD, Aanonsen L, Hargreaves KM. Hypoactivity of the spinal cannabinoid system results in NMDA-dependent hyperalgesia. J Neurosci. 1998;18(1):451–7. https://doi.org/10.1523/JNEUROSCI.18-01-00451.1998.
36. Mason DJ Jr, Lowe J, Welch SP. Cannabinoid modulation of dynorphin A: correlation to cannabinoid-induced antinociception. Eur J Pharmacol. 1999;378(3):237–48. https://doi.org/10.1016/s0014-2999(99)00479-3.
37. Hohmann AG, Tsou K, Walker JM. Cannabinoid suppression of noxious heat-evoked activity in wide dynamic range neurons in the lumbar dorsal horn of the rat. J Neurophysiol. 1999;81(2):575–83. https://doi.org/10.1152/jn.1999.81.2.575.
38. Jiang Y, Nie Y, Li Y, Zhang L. Association of cannabinoid type 1 receptor and fatty acid amide hydrolase genetic polymorphisms in Chinese patients with irritable bowel syndrome. J Gastroenterol Hepatol. 2014;29(6):1186–91. https://doi.org/10.1111/jgh.12513.
39. Cichewicz DL, McCarthy EA. Antinociceptive synergy between delta(9)-tetrahydrocannabinol and opioids after oral administration. J Pharmacol Exp Ther. 2003;304(3):1010–5. https://doi.org/10.1124/jpet.102.045575.
40. Williams J, Haller VL, Stevens DL, Welch SP. Decreased basal endogenous opioid levels in diabetic rodents: effects on morphine and delta-9-tetrahydrocannabinoid-induced antinociception. Eur J Pharmacol. 2008;584(1):78–86. https://doi.org/10.1016/j.ejphar.2007.12.035.
41. da Fonseca PD, Klein A, de Castro PA, da Fonseca Pacheco CM, de Francischi JN, Duarte ID. The mu-opioid receptor agonist morphine, but not agonists at delta- or kappa-opioid receptors, induces peripheral antinociception mediated by cannabinoid receptors. Br J Pharmacol. 2008;154(5):1143–9. https://doi.org/10.1038/bjp.2008.175.
42. Corchero J, Oliva JM, García-Lecumberri C, Martin S, Ambrosio E, Manzanares J. Repeated administration with Delta9-tetrahydrocannabinol regulates mu-opioid receptor density in the rat brain. J Psychopharmacol. 2004;18(1):54–8. https://doi.org/10.1177/0269881104040237.
43. Welch SP, Eads M. Synergistic interactions of endogenous opioids and cannabinoid systems. Brain Res. 1999;848(1–2):183–90. https://doi.org/10.1016/s0006-8993(99)01908-3.
44. Corchero J, Avila MA, Fuentes JA, Manzanares J. delta-9-Tetrahydrocannabinol increases prodynorphin and proenkephalin gene expression in the spinal cord of the rat. Life Sci. 1997;61(4):39–43. https://doi.org/10.1016/s0024-3205(97)00405-0.

45. Viganò D, Rubino T, Parolaro D. Molecular and cellular basis of cannabinoid and opioid interactions. Pharmacol Biochem Behav. 2005;81(2):360–8. https://doi.org/10.1016/j.pbb.2005.01.021.
46. Shamamian PE, Kwon DY, Oleru O, Seyidova N, Kim E, Genet S, Gyasi A, Wang CY, Henderson PW. The use of recreational cannabis among breast reduction patients: characteristics, complications, and immediate postoperative analgesic needs. Plast Surg (Oakv). 2024;30:22925503241276549. https://doi.org/10.1177/22925503241276549.
47. Maskal S, Foreman JM, Ellis RC, Phillips S, Messer N, Melland-Smith M, Beffa LRA, Petro CC, Prabhu AS, Rosen MJ, Miller BT. Cannabis smoking and abdominal wall reconstruction outcomes: a propensity score-matched analysis. Hernia. 2024;28(3):847–55. https://doi.org/10.1007/s10029-024-02976-3.

Central Nervous System: Pain, Movement, and Memory

11

11.1 Introduction to the ECS in the CNS

As previously stated, the best studied area of the ECS function is the central nervous system. As the CNS is a large and complicated structure, it will be divided into a number of sections based on the ECS's physiologic function. We will continue with the discussion of the ECS in pain perception, as well the ECS's role in movement and coordination, memory, sleep, cognition, and neuroprotection. We will also examine its role in a number of pathologic conditions including seizure disorders, migraine, and neurodegenerative and inflammatory diseases. Exploration of its role in emotional control and reward will be covered in a subsequent chapter focusing on psychiatric effects. In the chapters on central control of the CV system, respiration and GI motility, we detailed the presence of the ECS in the nuclei of the brain stem involved in autonomic control: such as the rostral ventrolateral nuclei that control sympathetic activity in blood pressure control, the solitary tract nucleus that controls sympathetic and parasympathetic baroreflexes, preBotzinger complex of the medulla that controls respiration, and the dorsal vagal complex that controls nausea and vomiting. We also saw the role of the ECS in the hypothalamus to influence appetite and metabolism in Chap. 4. As previously noted, CB1 receptor along with all of the other major components of the ECS is extensively expressed throughout the brain, while CB2 receptor is mostly limited to immune cells such as microglia and astrocytes. CB1 receptors are capable of regulating nearly all of the major neurotransmitters and modulate serotonergic, noradrenergic, dopaminergic, cholinergic, GABA, and glutaminergic neurons [1–6]. Many of the complex brain system discussed in this chapter are not completely understood, nor is the role of the ECS in that system. This leads to many instances of seemingly contradicting findings. The ECS's role as a feedback inhibitor lends itself to correcting seemingly contradicting extremes of neurotransmitter states, be they excess or deficit, as its main role is actually to restore balance, rather than always tipping the scale one way or the other. It is very important, therefore, not to always focus on the ECS inhibiting a

J.-P. H. Dedam et al., *Diseases of the Endocannabinoid System*,
https://doi.org/10.1007/978-3-031-87484-0_11

specific pathway or activating another, as in complicated systems like the brain, as we will see over and over again, it is very often doing both. Whether it does one or the other depends on which neurotransmitter happens to be in excess, the ECS will function by feedback inhibition to correct it. In these complicated systems, therefore, a dysfunctional ECS will be incapable of restoring balance and whatever neurotransmitter is in excess will remain that way. A primary disruption of the ECS in these same complicated systems would allow normal small fluctuations in neurotransmitters to grow into bigger ones, capable of pathologic states.

11.2 Pain

We now return to the discussion of pain perception and ECS, specifically focusing now on centrally mediated mechanisms. Early studies showed that injection of CB1 agonists into the periventricular areas of the brain had an analgesic effect in rats, specifically activating the dorsolateral periaqueductal gray matter and the dorsal raphe [7].

11.2.1 ECS in Periaqueductal Gray Matter

Subsequent study of the periaqueductal gray matter (PAG), a region known to be involved in pain modulation, showed endogenous increases in AEA here after painful stimuli. The PAG is activated by inputs arriving from the spinal dorsal horn carrying nerve impulses activated by painful stimuli and connects to other parts of the brain such as the amygdala and rostral medulla to elicit descending inhibition of pain transmission from the dorsal horn. Opiate receptors are also present here as well as in other parts of the brain pain modulation circuitry and perform the same function; however, they are only released with pain. Experiments have showed that CB1 blockage in this area of the brain increased pain sensitivity in the absence of an activating stimuli, indicating that the ECS plays a tonic role in pain inhibition [8]. Here again, the possibility of an ECS deficiency could result in the development of pain hyperalgesia and chronic pain syndrome. As has been mentioned in the previous chapter, human studies have shown that patients with chronic lower back pain have higher levels of circulating FAAH, which would in turn decrease available AEA [9]. The analgesic effect of CB1 in the PAG is mediated via inhibition of GABA and glutamatergic neurons [10]. Repeated administration of opiates leads to reduced inhibition of GABA neurons here, leading to tolerance, a significant clinical problem with the use of chronic opiates. One experiment demonstrated that opiate tolerance in the PAG resulted in increased ECS tone due to the upregulation of CB1 receptors. CB1 activation was then shown to result in greater analgesic effect of the opiate [11]. Given this finding, one might expect to see better pain relief than expected from co-administration of exogenous cannabinoids in patients who are already opiate tolerant. Some PAG glutamatergic neurons express both CB1 and TRPV1. Here, TRPV1 activation contributed to analgesic effects via glutamatergic neuron activation, while CB1 inhibition actually produced hyper-analgesia [12]. Orexin, a hormone released from the hypothalamus under stressful conditions and which also mediates wakefulness and feeding, was also found to have an analgesic effect in the PAG by stimulating the synthesis of 2-AG, which then inhibited GABA transmission via the mechanism described above [13].

11.2.2 ECS and Rostral Ventromedial Medulla

The ECS plays a similar role in another pain-modulating region of the brain: the rostral ventromedial medulla. Here, inhibition of glutamatergic neurons also activated descending analgesic effects [14]. The dorsal raphe, as mentioned above, also plays a role in central analgesia. In another sciatic nerve ligation model of neuropathic pain in rats, AEA levels were found to be elevated but not 2-AG. Increased ECS tone decreased mechanical and thermal hyperalgesia as well as increased serotonin output, with the dorsal raphe being a significant source of serotonergic neurons in the brain [15]. Further effects on the dorsal raphe and increased serotonergic tone by the ECS will be explored in the following chapter as this likely has significant effects on stress response, anxiety, and depression [16]. The same areas of the brain are also involved in stress-mediated analgesia. Injection of FAAH blockers increased stress-induced analgesia, and CB1 receptor antagonist blocked it in the PAG and rostral ventromedial medulla. TRPV1 receptors do not appear to play a role in this mechanism [17].

11.2.3 ECS in Limbic Structures

Cannabinoid receptors are expressed extensively in the limbic areas of the brain. They are in the hippocampus, amygdala, substantia nigra, and pars compacta and to a lesser degree in the globus pallidus and substantia nigra reticulata and are present both pre- and postsynaptically [18]. Blockade of CB1 receptors in the basolateral nucleus of the amygdala, the main threat perception center of the brain, also prevented stress-induced analgesia [19]. Activation of CB1 receptors in the substantia nigra also produced antinociceptive effects in mice by inhibition of substance P release, in addition to causing catalepsy [20]. CB1 receptors are present in the neocortex with highest densities in the prefrontal and cingulate cortex with lesser expression on the primary motor and sensory cortices [21]. Neocortical areas play a role in the affective-motivation dimension of pain; however, there are no direct studies looking at the ECS's role in central-mediated pain perception here [22].

Another central and peripheral mechanism of ECS pain modulation is the modulation of the release of calcitonin gene-related peptide (CGRP), which has been associated with somatic, visceral, neuropathic, and inflammatory pain. An experiment that focused on trigeminal nerve pain in rats showed that AEA prevented the release of CGRP both synaptically and postsynaptically. This was shown to decrease

NO release mediated by CGRP, which results in dural blood vessel dilation and headache [23].

11.3 Movement

Voluntary movement is prepared and initiated in the motor cortex of the brain. These neurons then project to the thalamus and to the spinal cord, where the signal is relayed to muscles to initiate action. Other brain regions, though, including the cerebellum and basal ganglia, are involved in fine-tuning these movements [24]. Numerous studies have demonstrated the effect of cannabis on coordination.

11.3.1 ECS in the Cerebellum

As mentioned above, CB1 receptors are present in parts of the basal ganglia involved in movement but also in high amounts on the Purkinje cells of the cerebellum, which are essential for the coordination of movement. Eye blink conditioning, an experiment that tests cerebellum-dependent learning, has been shown to be severely impaired in CB1 knockout mice [25]. Excessive activation of the ECS can also disrupt cerebellar learning, as has been demonstrated in chronic cannabis users who also demonstrate impaired eye blink conditioning and that this disruption can persist after cessation as well [26]. Purkinje cells are the only efferent neurons of the cerebellum and project to the deep cerebellar nucleus where they release GABA to inhibit neurons that project to other sites of the brain to coordinate movement and other functions such as emotion [27]. In the cerebellum, the ECS is intimately involved in long-term synaptic plasticity, when the signal across the synapsis changes as a result of activity in the synapsis. Synaptic plasticity is the cornerstone of learned complex behaviors such as eating or driving a car. This plasticity occurs at the granule cell-Purkinje cell synapse, which is regulated but the ECS via retrograde inhibition [28]. Purkinje cells produce and release 2-AG in response to depolarization or activation of glutamate receptors by glutamate from granule cells [29]. Activation of CB1 receptors then reduces neurotransmitter release from all Purkinje cell synapses [30]. Inhibitory interneurons of the cerebellum and stellate and basket cells can also release endocannabinoids that affect Purkinje cell's neurotransmitter release [31].

11.3.2 ECS in the Spinocerebellar Ataxias

Postmortem studies of patients with spinocerebellar ataxias, including Huntington's disease, which are characterized by loss of balance and motor function due to degeneration of cerebellar connections, have shown increased expression of FAAH and MAGL in the cerebellum, especially in Purkinje cells [32, 33]. CB1 and CB2 receptors are also present in increased amounts on the same cells as presented in a separate study [32, 33]. This disease process is known to involve excitotoxicity, local inflammation, and oxidative stress, leading to death of Purkinje cells. The ECS is part of the body's defense against all of these, which is to be discussed in the neuroprotection section, and any decrease in ECS function would certainly accelerate cell death and disease progression. However, disruption of the ECS may also contribute to the symptomatology in the form of Purkinje cell dysfunction.

11.3.3 ECS in the Globus Pallidus and Huntington's Disease

Another site of motor control in the body is the globus pallidus. Here, inhibitory GABAergic inputs from the striatum are integrated with excitatory glutamatergic inputs from the subthalamic nucleus, neocortex and thalamus [34]. PET scan studies have demonstrated that the globus pallidus in humans has the highest binding activity of CB1 aside from the cortex itself and that this activity declines with normal aging [35]. Immunohistological studies have further localized these receptors to the presynaptic GABA inputs from the striatum mentioned above [36]. The ECS in the globus pallidus acts in the classic feedback inhibition loop: decreasing inhibitory inputs. As the main output for the globus pallidus is inhibitory signals to the thalamus that coordinate movement, the activation of the ECS here results in reduced movement or catalepsy in mice studies [37]. The loss of these GABA neurons is part of the pathophysiology of Huntington's disease, and the loss of CB1 receptors in the globus pallidus is seen in the akinetic late stage of the disease in rats [38, 39]. Early-stage Huntington's also has loss of CB1 receptors but is characterized by uncontrollable hyperkinetic movements. Studies in rats at this stage showed improvement in symptoms after increased ECS tone [38, 39]. Balancing the inhibitory GABA inputs is the presence of CB1 receptors in the excitatory glutamatergic inputs from the subthalamic nucleus, neocortex, and thalamus, which have also been shown to be inhibited via the same feedback inhibition mechanism. Activation of CB1 receptors here resulted in cataplexy or decreased movement seen in administration of high doses of THC to animals, also known colloquially in humans as couch lock [40].

11.3.4 ECS in the Striatum

The striatum is also involved in procedural learning, defined as the acquisition of skills and the practice that refines their performance. ATT repeat polymorphisms in an untranslated portion of the CR1 genes have been associated with addiction, with greater repeats leading to greater risk, although the actual effect on gene expression is still unknown [41]. The number of AAT repeats has also been shown to affect procedural learning, indicating the role of the ECS here as well, although the exact mechanism remains poorly understood [42]. CB1 receptor expression has also shown to be decreased in postmortem studies of Parkinson's patients [43]. However, there exists currently conflicting studies in animals showing benefit of CB1 agonist, as well as CB1 antagonism in models of this disease.

11.4 Memory and the ECS of the Hippocampus

Human episodic memories, the memories formed that remind us of what has happened previously, where, why, and what emotions were associated with that memory, are formed in the hippocampus. The hippocampus is further divided into several regions, e.g., the dentate gyrus, where sensory information enters from the cortex and then information passes to CA3 region via mossy fibers and then to CA1 via Schaffer's collaterals. CA1 then projects to the subiculum, which contacts many different areas of the brain, including the cortex. The underlying cellular mechanism of memory is long-term potentiation (LTP) where the strength of the signal across a particular synapsis is increased with further activation. To begin this process, the neurotransmitter glutamate is released across the synapsis, with depolarization of the postsynaptic cell. The rise in intracellular calcium leads to synaptic modifications, thereby strengthening the connection between the two neurons and forming a memory trace. CB1 receptors are also extensively expressed here, located primarily in GABAergic interneurons but also in glutamatergic, cholinergic, and dopaminergic neurons as well as astrocytes. Glutamatergic principal cells have few CB1 receptors but express large amounts of DAGL, the enzymes responsible for producing 2-AG [44]. Cholinergic projections are largely responsible for spatial orientation memory generation and, in various experiments, have shown impairment of spatial navigation in rats after CB1-dependent decrease in acetylcholine release [45].

The first stage of memory generation is encoding, where information is initially learned. Cannabinoids have long been known to impair this by decreasing LTP in the CA1 region. Chronic administration has a different effect, with the first 5 days showing impairment in spatial memory tasks but then becoming indistinguishable from controls with subsequent dosing [46]. Chronic administration in older rats actually produced improved performance, via a CB1-dependent decrease in microglial cells, which mediate inflammation in the brain [47]. This may also be related to CB1 activation of hippocampal cell neurogenesis [48]. CB1 blockade has mixed effects on memory. Studies looking at the development of contextual fear memories, memories associated with a specific context such as water emersion, showed enhanced fear memory generation with CB1 blockage and reduction in fear memories with stimulation via blockage of LTP of GABA interneurons [49]. This may play a role in the pathophysiology of PTSD, which we will revisit in sections on consolidation and extinction of memory. Blockage of CB1 receptors, however, also impaired spatial memory, demonstrating that memory generation does still require some endocannabinoid activity [50].

The next stage of memory generation is consolidation, when newly encoded memories are retained. Administration of CB1 agonist immediately following a training phase of memory tasks leads to learned memory deficits in rats [51]. CB1 blockers administered at the same time improved later performances of the learned task. Consolidation of contextual fear memories is dependent on a connection between the subiculum of the hippocampus and the amygdala. Fear memory consolidation was noted to enhance in a study of mice with genetic deletion of CB1 receptor selectively in the amygdala [52]. Consolidated memories must then be retrieved. Administration of CB1 agonist prior to or during a previously learned task results in impaired performance, indicating worsened retrieval. Glucocorticoids have long been known to impair retrieval of contextual fear memories, but so can cannabinoids. Glucocorticoids are able to increase cannabinoid levels in the hippocampus to impair retrieval, which is also dependent on the activity of noradrenergic neurons projecting from the locus coeruleus [53]. CB1 activation also facilitates extinction of memories, or forgetting, increasing the rate at which rats forget how to perform a particular task [54].

Hemopressin, a pepcan and inverse CB1 agonist, has a very different effect on memory from other endocannabinoids. Studies in mice have revealed improvement in memory tasks, specifically in Alzheimer models [55].

11.5 ECS and PTSD

PTSD is associated with increased activity in the amygdala. Endocannabinoid function is again necessary in the amygdala for fear extinction. Fear extinction is associated with increases in AEA in the amygdala, and blocking CB1 receptor here delays fear extinction [56]. Posttraumatic stress disorder is characterized by slower or nonexistent extinction of fear responses. Circulating endocannabinoids measured in survivors of 9/11 showed that elevated levels of AEA protected against the development of PTSD and low levels of 2-AG were associated with higher rates of PTSD [57]. This demonstrates that an endocannabinoid deficiency could be a significant risk factor in the development of PTSD, if not the outright cause. FAAH polymorphism C385A/rs324420, which causes impaired function and increased AEA levels in the body, has been shown to confer more successful memory extinction than subjects without it [58]. Subsequent studies using functional MRI in subjects with the same polymorphism showed less amygdala and dorsal anterior cingulate cortex activation with increased hippocampus activation during memory extinction [59].

As seen above, the endocannabinoid system plays an active role at every level of memory formation with acute administration of cannabinoids largely impairing memory formation, although this effect disappears with chronic administration. While this might not seem like positive, forgetting can be extremely important, especially in the context of diseases like PTSD. PTSD sufferers have long used cannabis to manage their symptoms after trauma, but as seen above, modification of the ECS may actually have a role to play in the prevention of the condition as well.

References

1. Häring M, Marsicano G, Lutz B, Monory K. Identification of the cannabinoid receptor type 1 in serotonergic cells of raphe nuclei in mice. Neuroscience. 2007;146(3):1212–9. https://doi.org/10.1016/j.neuroscience.2007.02.021.
2. Oropeza VC, Mackie K, Van Bockstaele EJ. Cannabinoid receptors are localized to noradrenergic axon terminals in the rat frontal cortex. Brain Res. 2007;1127(1):36–44. https://doi.org/10.1016/j.brainres.2006.09.110.
3. Rodríguez De Fonseca F, Gorriti MA, Bilbao A, Escuredo L, García-Segura LM, Piomelli D, Navarro M. Role of the endogenous cannabinoid system as a modulator of dopamine transmission: implications for Parkinson's disease and schizophrenia. Neurotox Res. 2001;3(1):23–35. https://doi.org/10.1007/BF03033228.
4. Nyíri G, Szabadits E, Cserép C, Mackie K, Shigemoto R, Freund TF. GABAB and CB1 cannabinoid receptor expression identifies two types of septal cholinergic neurons. Eur J Neurosci. 2005;21(11):3034–42. https://doi.org/10.1111/j.1460-9568.2005.04146.x.
5. Hájos N, Katona I, Naiem SS, MacKie K, Ledent C, Mody I, Freund TF. Cannabinoids inhibit hippocampal GABAergic transmission and network oscillations. Eur J Neurosci. 2000;12(9):3239–49. https://doi.org/10.1046/j.1460-9568.2000.00217.x.
6. Doherty J, Dingledine R. Functional interactions between cannabinoid and metabotropic glutamate receptors in the central nervous system. Curr Opin Pharmacol. 2003;3(1):46–53. https://doi.org/10.1016/s1471-4892(02)00014-0.
7. Martin WJ, Patrick SL, Coffin PO, Tsou K, Walker JM. An examination of the central sites of action of cannabinoid-induced antinociception in the rat. Life Sci. 1995;56(23–24):2103–9. https://doi.org/10.1016/0024-3205(95)00195-c.
8. Walker JM, Huang SM, Strangman NM, Tsou K, Sañudo-Peña MC. Pain modulation by release of the endogenous cannabinoid anandamide. Proc Natl Acad Sci USA. 1999;96(21):12198–203. https://doi.org/10.1073/pnas.96.21.12198.
9. Ramesh D, D'Agata A, Starkweather AR, Young EE. Contribution of endocannabinoid gene expression and genotype on low Back pain susceptibility and chronicity. Clin J Pain. 2018;34(1):8–14. https://doi.org/10.1097/AJP.0000000000000508.
10. Vaughan CW, Connor M, Bagley EE, Christie MJ. Actions of cannabinoids on membrane properties and synaptic transmission in rat periaqueductal gray neurons in vitro. Mol Pharmacol. 2000;Feb;57(2):288–95. PMID: 10648638.
11. Wilson-Poe AR, Lau BK, Vaughan CW. Repeated morphine treatment alters cannabinoid modulation of GABAergic synaptic transmission within the rat periaqueductal grey. Br J Pharmacol. 2015;Jan;172(2):681–90. https://doi.org/10.1111/bph.12809. Epub 2014 Sep 5. PMID: 24916363; PMCID: PMC4292978.
12. Kawahara H, Drew GM, Christie MJ, Vaughan CW. Inhibition of fatty acid amide hydrolase unmasks CB1 receptor and TRPV1 channel-mediated modulation of glutamatergic synaptic transmission in midbrain periaqueductal grey. Br J Pharmacol. 2011;163(6):1214–22. https://doi.org/10.1111/j.1476-5381.2010.01157.x.
13. Ho YC, Lee HJ, Tung LW, Liao YY, Fu SY, Teng SF, Liao HT, Mackie K, Chiou LC. Activation of orexin 1 receptors in the periaqueductal gray of male rats leads to antinociception via retrograde endocannabinoid (2-arachidonoylglycerol)-induced disinhibition. J Neurosci. 2011;31(41):14600–10. https://doi.org/10.1523/JNEUROSCI.2671-11.2011.
14. de Novellis V, Mariani L, Palazzo E, Vita D, Marabese I, Scafuro M, Rossi F, Maione S. Periaqueductal grey CB1 cannabinoid and metabotropic glutamate subtype 5 receptors modulate changes in rostral ventromedial medulla neuronal activities induced by subcutaneous formalin in the rat. Neuroscience. 2005;134(1):269–81. https://doi.org/10.1016/j.neuroscience.2005.03.014.
15. Palazzo E, de Novellis V, Petrosino S, Marabese I, Vita D, Giordano C, Di Marzo V, Mangoni GS, Rossi F, Maione S. Neuropathic pain and the endocannabinoid system in the dorsal raphe:

pharmacological treatment and interactions with the serotonergic system. Eur J Neurosci. 2006;24(7):2011–20. https://doi.org/10.1111/j.1460-9568.2006.05086.x.
16. Haj-Dahmane S, Shen RY. Modulation of the serotonin system by endocannabinoid signaling. Neuropharmacology. 2011;61(3):414–20. https://doi.org/10.1016/j.neuropharm.2011.02.016.
17. Suplita RL 2nd, Farthing JN, Gutierrez T, Hohmann AG. Inhibition of fatty-acid amide hydrolase enhances cannabinoid stress-induced analgesia: sites of action in the dorsolateral periaqueductal gray and rostral ventromedial medulla. Neuropharmacology. 2005;49(8):1201–9. https://doi.org/10.1016/j.neuropharm.2005.07.007.
18. Ong WY, Mackie K. A light and electron microscopic study of the CB1 cannabinoid receptor in primate brain. Neuroscience. 1999;92(4):1177–91. https://doi.org/10.1016/s0306-4522(99)00025-1.
19. Connell K, Bolton N, Olsen D, Piomelli D, Hohmann AG. Role of the basolateral nucleus of the amygdala in endocannabinoid-mediated stress-induced analgesia. Neurosci Lett. 2006;397(3):180–4. https://doi.org/10.1016/j.neulet.2005.12.008.
20. Soria-Gomez E, Pagano Zottola AC, Mariani Y, Desprez T, Barresi M, Bonilla-Del Río I, Muguruza C, Le Bon-Jego M, Julio-Kalajzić F, Flynn R, Terral G, Fernández-Moncada I, Robin LM, Oliveira da Cruz JF, Corinti S, Amer YO, Goncalves J, Varilh M, Cannich A, Redon B, Zhao Z, Lesté-Lasserre T, Vincent P, Tolentino-Cortes T, Busquets-García A, Puente N, Bains JS, Hebert-Chatelain E, Barreda-Gómez G, Chaouloff F, Lohman AW, Callado LF, Grandes P, Baufreton J, Marsicano G, Bellocchio L. Subcellular specificity of cannabinoid effects in striatonigral circuits. Neuron. 2021;109(9):1513–26. https://doi.org/10.1016/j.neuron.2021.03.007.
21. Eggan SM, Lewis DA. Immunocytochemical distribution of the cannabinoid CB1 receptor in the primate neocortex: a regional and laminar analysis. Cereb Cortex. 2007;17(1):175–91. https://doi.org/10.1093/cercor/bhj136.
22. Rea K, Roche M, Finn DP. Supraspinal modulation of pain by cannabinoids: the role of GABA and glutamate. Br J Pharmacol. 2007;152(5):633–48. https://doi.org/10.1038/sj.bjp.0707440.
23. Akerman S, Kaube H, Goadsby PJ. Anandamide is able to inhibit trigeminal neurons using an in vivo model of trigeminovascular-mediated nociception. J Pharmacol Exp Ther. 2004;309(1):56–63. https://doi.org/10.1124/jpet.103.059808.
24. Rocha GS, Freire MAM, Britto AM, Paiva KM, Oliveira RF, Fonseca IAT, Araújo DP, Oliveira LC, Guzen FP, Morais PLAG, Cavalcanti JRLP. Basal ganglia for beginners: the basic concepts you need to know and their role in movement control. Front Syst Neurosci. 2023;17:1242929. https://doi.org/10.3389/fnsys.2023.1242929.
25. Kishimoto Y, Kano M. Endogenous cannabinoid signaling through the CB1 receptor is essential for cerebellum-dependent discrete motor learning. J Neurosci. 2006;26(34):8829–37. https://doi.org/10.1523/JNEUROSCI.1236-06.2006.
26. Steinmetz AB, Edwards CR, Vollmer JM, Erickson MA, O'Donnell BF, Hetrick WP, Skosnik PD. Examining the effects of former cannabis use on cerebellum-dependent eyeblink conditioning in humans. Psychopharmacology. 2012;221(1):133–41. https://doi.org/10.1007/s00213-011-2556-1.
27. Goodlett CR, Mittleman G. Conn's translational neuroscience; 2017.
28. Marcaggi P. Cerebellar endocannabinoids: retrograde signaling from purkinje cells. Cerebellum. 2015;Jun;14(3):341–53. https://doi.org/10.1007/s12311-014-0629-5. PMID: 25520276.
29. Safo PK, Cravatt BF, Regehr WG. Retrograde endocannabinoid signaling in the cerebellar cortex. Cerebellum. 2006;5(2):134–45. https://doi.org/10.1080/14734220600791477.
30. Takahashi KA, Linden DJ. Cannabinoid receptor modulation of synapses received by cerebellar Purkinje cells. J Neurophysiol. 2000;83(3):1167–80. https://doi.org/10.1152/jn.2000.83.3.1167.
31. Beierlein M, Regehr WG. Local interneurons regulate synaptic strength by retrograde release of endocannabinoids. J Neurosci. 2006;26(39):9935–43. https://doi.org/10.1523/JNEUROSCI.0958-06.2006.
32. Rodríguez-Cueto C, Benito C, Romero J, Hernández-Gálvez M, Gómez-Ruiz M, Fernández-Ruiz J. Endocannabinoid-hydrolysing enzymes in the post-mortem cerebellum of humans

affected by hereditary autosomal dominant ataxias. Pathobiology. 2014a;81(3):149–59. https://doi.org/10.1159/000358127.

33. Rodríguez-Cueto C, Benito C, Fernández-Ruiz J, Romero J, Hernández-Gálvez M, Gómez-Ruiz M. Changes in CB(1) and CB(2) receptors in the post-mortem cerebellum of humans affected by spinocerebellar ataxias. Br J Pharmacol. 2014b;171(6):1472–89. https://doi.org/10.1111/bph.12283.
34. Chen XY, Xue Y, Chen H, Chen L. The globus pallidus as a target for neuropeptides and endocannabinoids participating in central activities. Peptides. 2020;124:170210. https://doi.org/10.1016/j.peptides.2019.170210.
35. Wong DF, Kuwabara H, Horti AG, Raymont V, Brasic J, Guevara M, Ye W, Dannals RF, Ravert HT, Nandi A, Rahmim A, Ming JE, Grachev I, Roy C, Cascella N. Quantification of cerebral cannabinoid receptors subtype 1 (CB1) in healthy subjects and schizophrenia by the novel PET radioligand [11C] OMAR. NeuroImage. 2010;52(4):1505–13. https://doi.org/10.1016/j.neuroimage.2010.04.034.
36. Davis MI, et al. The cannabinoid-1 receptor is abundantly expressed in striatal striosomes and striosome-dendron bouquets of the substantia nigra. PLoS One. 2018;13(2):e0191436.
37. Wallmichrath I, Szabo B. Cannabinoids inhibit striatonigral GABAergic neurotransmission in the mouse. Neuroscience. 2002;113(3):671–82. https://doi.org/10.1016/s0306-4522(02)00109-4.
38. Lastres-Becker I, Gómez M, De Miguel R, Ramos JA, Fernández-Ruiz J. Loss of cannabinoid CB(1) receptors in the basal ganglia in the late akinetic phase of rats with experimental Huntington's disease. Neurotox Res. 2002a;4(7–8):601–8. https://doi.org/10.1080/1029842029003051 4.
39. Lastres-Becker I, Hansen HH, Berrendero F, De Miguel R, Pérez-Rosado A, Manzanares J, Ramos JA, Fernández-Ruiz J. Alleviation of motor hyperactivity and neurochemical deficits by endocannabinoid uptake inhibition in a rat model of Huntington's disease. Synapse. 2002b;44(1):23–35. https://doi.org/10.1002/syn.10054.
40. Freiman I, Szabo B. Cannabinoids depress excitatory neurotransmission between the subthalamic nucleus and the globus pallidus. Neuroscience. 2005;133(1):305–13. https://doi.org/10.1016/j.neuroscience.2005.01.058.
41. Benyamina A, Kebir O, Blecha L, Reynaud M, Krebs MO. CNR1 gene polymorphisms in addictive disorders: a systematic review and a meta-analysis. Addict Biol. 2011;16(1):1–6. https://doi.org/10.1111/j.1369-1600.2009.00198.x.
42. Ruiz-Contreras AE, Delgado-Herrera M, García-Vaca PA, Almeida-Rosas GA, Soria-Rodríguez G, Soriano-Bautista A, Cadena-Valencia J, Bazán-Frías JR, Gómez-López N, Espejel-Núñez A, Vadillo-Ortega F, Carrillo-Sánchez K, Verdín-Reyes JC, March-Mifsut S, Méndez-Díaz M, Prospéro-García O. Involvement of the AATn polymorphism of the CNR1 gene in the efficiency of procedural learning in humans. Neurosci Lett. 2011;494(3):202–6. https://doi.org/10.1016/j.neulet.2011.03.013.
43. Hurley MJ, Mash DC, Jenner P. Expression of cannabinoid CB1 receptor mRNA in basal ganglia of normal and parkinsonian human brain. J Neural Transm (Vienna). 2003;110(11):1279–88. https://doi.org/10.1007/s00702-003-0033-7.
44. Figueiredo A, Cheer JF. Endocannabinoid regulation of hippocampus-dependent memory. Exp Neurol. 2023;364:114384. https://doi.org/10.1016/j.expneurol.2023.114384.
45. Gessa GL, Mascia MS, Casu MA, Carta G. Inhibition of hippocampal acetylcholine release by cannabinoids: reversal by SR 141716A. Eur J Pharmacol. 1997;327(1):R1–2. https://doi.org/10.1016/s0014-2999(97)89683-5.
46. Hill MN, Froc DJ, Fox CJ, Gorzalka BB, Christie BR. Prolonged cannabinoid treatment results in spatial working memory deficits and impaired long-term potentiation in the CA1 region of the hippocampus in vivo. Eur J Neurosci. 2004;20(3):859–63. https://doi.org/10.1111/j.1460-9568.2004.03522.x.
47. Marchalant Y, Cerbai F, Brothers HM, Wenk GL. Cannabinoid receptor stimulation is anti-inflammatory and improves memory in old rats. Neurobiol Aging. 2008;29(12):1894–901. https://doi.org/10.1016/j.neurobiolaging.2007.04.028.

48. Andres-Mach M, Haratym-Maj A, Zagaja M, Rola R, Maj M, Chrościńska-Krawczyk M, Luszczki JJ. ACEA (a highly selective cannabinoid CB1 receptor agonist) stimulates hippocampal neurogenesis in mice treated with antiepileptic drugs. Brain Res. 2015;1624:86–94. https://doi.org/10.1016/j.brainres.2015.07.028.
49. Lin QS, Yang Q, Liu DD, Sun Z, Dang H, Liang J, Wang YX, Chen J, Li ST. Hippocampal endocannabinoids play an important role in induction of long-term potentiation and regulation of contextual fear memory formation. Brain Res Bull. 2011;86(3–4):139–45. https://doi.org/10.1016/j.brainresbull.2011.07.011.
50. Abush H, Akirav I. Cannabinoids modulate hippocampal memory and plasticity. Hippocampus. 2010;20(10):1126–38. https://doi.org/10.1002/hipo.20711.
51. Hasanein P, Teimuri FM. Effects of URB597 as an inhibitor of fatty acid amide hydrolase on WIN55, 212-2-induced learning and memory deficits in rats. Pharmacol Biochem Behav. 2015;131:130–5. https://doi.org/10.1016/j.pbb.2015.02.007.
52. Kondev V, Najeed M, Yasmin F, Morgan A, Loomba N, Johnson K, Adank DN, Dong A, Delpire E, Li Y, Winder D, Grueter BA, Patel S. Endocannabinoid release at ventral hippocampal-amygdala synapses regulates stress-induced behavioral adaptation. Cell Rep. 2023;42(9):113027. https://doi.org/10.1016/j.celrep.2023.113027.
53. Atsak P, Hauer D, Campolongo P, Schelling G, McGaugh JL, Roozendaal B. Glucocorticoids interact with the hippocampal endocannabinoid system in impairing retrieval of contextual fear memory. Proc Natl Acad Sci USA. 2012;109(9):3504–9. https://doi.org/10.1073/pnas.1200742109.
54. Varvel SA, Wise LE, Niyuhire F, Cravatt BF, Lichtman AH. Inhibition of fatty-acid amide hydrolase accelerates acquisition and extinction rates in a spatial memory task. Neuropsychopharmacology. 2007;32(5):1032–41. https://doi.org/10.1038/sj.npp.1301224.
55. Zhang RS, He Z, Jin WD, Wang R. Effects of the cannabinoid 1 receptor peptide ligands hemopressin, (m)RVD-hemopressin(α) and (m)VD-hemopressin(α) on memory in novel object and object location recognition tasks in normal young and Aβ1-42-treated mice. Neurobiol Learn Mem. 2016;134:264–74. https://doi.org/10.1016/j.nlm.2016.07.030.
56. Adhikari A. Endocannabinoids modulate fear extinction controlled by a cortical-amygdala projection. Neuron. 2023;111(19):2948–50. https://doi.org/10.1016/j.neuron.2023.09.011.
57. Hill MN, Bierer LM, Makotkine I, Golier JA, Galea S, McEwen BS, Hillard CJ, Yehuda R. Reductions in circulating endocannabinoid levels in individuals with post-traumatic stress disorder following exposure to the world trade center attacks. Psychoneuroendocrinology. 2013;38(12):2952–61. https://doi.org/10.1016/j.psyneuen.2013.08.004.
58. Spohrs J, Ulrich M, Grön G, Plener PL, Abler B. FAAH polymorphism (rs324420) modulates extinction recall in healthy humans: an fMRI study. Eur Arch Psychiatry Clin Neurosci. 2022;272(8):1495–504. https://doi.org/10.1007/s00406-021-01367-4.
59. Zabik NL, Iadipaolo AS, Marusak HA, Peters C, Burghardt K, Rabinak CA. A common genetic variant in fatty acid amide hydrolase is linked to alterations in fear extinction neural circuitry in a racially diverse, nonclinical sample of adults. J Neurosci Res. 2022;100(3):744–61. https://doi.org/10.1002/jnr.24860.

Central Nervous System: Sleep, Cognition and Neuroprotection

12

12.1 Sleep

Few physiologic processes are as undervalued as sleep. Study after study has shown it to be a cornerstone of so many other vital functions of the human body such as immunity and mental health. Cannabinoids have long been known to have a hypnotic effect, and now there is beginning to be some understanding of how the ECS is involved in regulating this important process. The sleep-wake cycle is complex and not fully understood in its own. The various stages are characterized by specific brainwaves seen in EEG and associated with varying degrees of arousal. We will first review the various stages, discussing the parts of the brain responsible each, and then see how the ECS plays a part.

Our waking state, characterized mostly by alpha and beta waves(8–20 Hz frequencies), is controlled by neurons of the lateral and posterior thalamus: producing neurotransmitters like orexin and histamine, the locus coeruleus that releases noradrenaline, the dorsal raphe nucleus, and the pons. Non-rapid eye movement sleep, which is further divided into three phases, i.e., lightest sleep, light sleep, and deep sleep, is characterized by slower frequency waves called delta waves (0.5–4 Hz). It is controlled by the hypothalamus and mediated by the neurotransmitter GABA, adenosine, and PGD2, a prostaglandin. It is vital for consolidation of memories. Rapid eye movement (REM) sleep is mostly beta wave dominated, and this is where our remembered dreams occur. It is controlled by the pontis oralis nuclei as well as the thalamic and brain stem structures using the neurotransmitters glutamate, acetylcholine, and the melanin-concentrating hormone [1]. While there are a number of stages, this does mean to imply that a person progresses in an orderly fashion from one stage to the next but rather, cycles through them continuously even under normal conditions. The coordination of all stages is based largely on our circadian rhythm, which is in turn a result of the Earth's rotation around its axis in relation to the Sun.

J.-P. H. Dedam et al., *Diseases of the Endocannabinoid System*,
https://doi.org/10.1007/978-3-031-87484-0_12

The circadian rhythm causes changes in the levels of endocannabinoids and its receptors in the brain as well. CB1 receptor concentrations in rats were shown to be highest around noon and lowest at midnight. CB2 receptor concentrations did not change over the course of the day [2]. 2-AG levels in human studies increase throughout the morning, peaking in early afternoon, and are at this point as high as three times greater than their levels at their nocturnal nadir [3]. AEA is biphasic in humans with a peak during light sleep and another peak in the midafternoon with a midmorning nadir in between. The overall delta in AEA levels, however, is smaller overall than 2-AG. Sleep restriction has been found to increase the 2-AG peak in the midafternoon but does not affect AEA levels [4].

OEA, a related endocannabinoid -like signaling lipid that does not bind well to the CB receptors, was found to decrease feeding and increase wakefulness in rats, possibly via binding of PPAR alpha [5]. Significant increases in OEA have been found also in patients with sleep apnea and thought to be a mechanism to increase wakefulness [6].

AEA was found in rats to promote non-REM sleep, apparently by increasing adenosine levels via CB1 activation [7]. AEA has also been shown in rats to increase REM sleep via CB1 receptor [8]. 2-AG promotes REM sleep too, specifically through interaction with melanin-concentrating hormone neurons of the lateral hypothalamus [9]. A meta-analysis of numerous human studies of dronabinol and nabilone did show improvement in sleep and lessening of nightmares in PTSD patients, although findings were inconsistent across numerous studies. Use of Sativex in human studies, which is a combination a CBD/THC musical spray, also improved sleep quality, although most of these studies were specifically end pointed for pain not sleep [10]. EEGs of human subjects given THC often show facilitation of falling asleep and increase in non-REM deep sleep but reduction in REM sleep [11]. This might explain the improvements seen in PTSD patients noted above and why this is often used by this patient population. Acute withdrawal from cannabis has also long been known to produce sleep difficulty [12]. This would indicate that the disruption in the ECS that occurs during the withdrawal affects sleep, strengthening the argument for a role of the ECS in regulating sleep. As we have seen above, normal sleep physiology is dependent on numerous signals across many brain structures, with the ECS being one of them. Insomnia is definitely due to a combination of many different factors, but given the role of the ECS system in sleep, at least some portion of this disease state is likely due to endocannabinoid deficiency syndrome as clearly demonstrated acute cannabis withdrawal, where the body suffers from a relative ECS deficiency. Given the ready availability of cannabis, it is no surprise that many insomnia sufferers also turn to it for help in sleeping.

12.2 Cognition

The consciousness altering effects of cannabis have been used for millennia by humans. Understanding how cannabis interacts with the ECS to affect our higher cognitive functions is only just beginning to be understood. One meta-analysis of cannabis users showed small magnitude deficits in executive function, learning and memory, and global function, with moderate deficits in decision-making, indicating the role of the ECS in all of these cognitive functions [13]. Other systemic reviews have found that impairments in memory and attention are the most significant cognitive impairments [14]. As previously mentioned, human cortices contain high densities of cannabinoid receptors in the forebrain specifically, and this distribution is shared across mammalian species [15].

12.2.1 ECS in the Default Mode Network

Goal-directed behavior is an important executive function that is often disrupted in the disease states like schizophrenia and ADHD, diseases known to be affected by the ECS. Goal-directed behavior has been shown to be associated with reduced activity in the default mode network (DMN). The DMN includes the medial prefrontal cortex, posterior cingulate gyrus, precuneus, and angular gyrus. It is activated when we daydream, think about ourselves, think about others, remember the past, and plan the future. Patients with schizophrenia and ADHD are often unable to deactivate the DMN, and this results in executive function impairments. Functional MRI was used in a study of healthy adults who were given THC and then underwent tasks testing executive function. Similar to patients with schizophrenia and ADHD, normal adults given THC were prevented from deactivating the DMN, which predictably interfered in goal-directed behavior [16]. While the ECS is known to be disrupted in both of these conditions, it does appear that it is the ECS that is primarily driving either of these disease states. The ECS does, however, as evidenced by cannabis intoxication seem to promote the DMN, promoting daydreaming, contemplation of oneself and others. While this would certainly interfere with "taking care of business," there remains an obvious draw as these activities are creative, pleasurable, and capable of giving value and meaning to our lives. In another neurologic condition with impaired executive function, i.e., traumatic brain injuries, non-goal-directed nighttime hyperactivity was associated with low levels of AEA [17]. Schizophrenia, discussed above, is generally known to have high levels of ECS activation. Neither elevated nor persistently low ECS function appears to be the answer to improve concentration.

12.2.2 ECS and the Salience Network

Similar MRI studies have also shown that THC is capable of reducing connectivity in the salience network [18]. The salience network consists of the cingulate gyrus and insular cortex as well as a number of other structures, and while its exact role is not fully understood, it is involved in detecting which of the many stimuli bombarding out brains is important enough to focus on. Disruption of this network is known to occur in the socially inappropriate behavior of frontotemporal dementia as well as many other neuropsychiatric diseases [19]. In the previously mentioned study, attention itself is a very complicated emergent property of many interacting brain networks and systems, ultimately coordinated by the prefrontal cortex [20]. Many of the component systems as well as the prefrontal cortex have CB1 receptors as previously discussed. The exact role of the ECS here is not understood, although the findings above are beginning to shed light on the known effects of THC on attention. The study also showed patients with reduction in the newly described hippocampus-midbrain-striatum network with THC. Disruptions in the network have also been seen in patients with schizophrenia [21].

12.2.3 ECS, Error Awareness and ADHD

Another important aspect of higher executive function is error awareness. Diminished capacity for this ability has been linked to clinical symptoms of delusions and loss of insight. The anterior cingulate gyrus is necessary for this function. Chronic cannabis users have been shown to demonstrate decreased error awareness associated with hypoactivity in the anterior cingulate and insular cortex [22]. The anterior cingulate gyrus is also known to be associated with impulsivity, specifically in the context of ADHD. One the cardinal features of ADHD is impulsivity [23]. Higher levels of glutamate and lower levels of GABA are associated with impulsivity in ADHD in human MR spectroscopy [24]. The ECS is capable of modulating both of these receptors, although unfortunately there do not appear to be any studies looking at the location of CB1 receptors on specific neurons of the cingulate gyrus other than them being present presynaptically on inhibitory cholecystokinin interneurons of mice. The blockade of these CB1 receptors lead to increased social isolation in mice [25]. A SNP in the gene encoding a CB1 receptor, rs2023239, which appears to confer increased activity, is associated with higher impulsivity in ADHD patients [26]. Many ADHD patients are known to self-medicate with cannabis however and one human study using Sativex, a combination of THC/CBD submucosal spray, showed improvement in hyperactivity/impulsivity [27]. ADHD likely represents a more complicated dysregulation in the ECS, rather than simply a pure deficiency or excess and probably also represents more than one common etiology, especially given the fact that the exact role of dopamine in this condition is still a matter of debate. The above findings of improvement in symptoms with Sativex are likely helped by the use of a medication with CBD as well as THC, which is known to modulate the effects of THC by allosterically limiting the activation of the CB1 receptor [28]. This prevents overstimulation of CB1 receptors in the brain, which can result in a variety of other cognitive deficits detailed above. In the mice model of ADHD, where the dopamine transporter gene is knocked out, resulting in a hyper-dopaminergic state which mimics some symptoms of ADHD (although some animal models show a hypo-dopaminergic state in the basal ganglia, resulting in disease presentation). AEA was found to be reduced in these mice and administration of CB1 agonist improved hyperlocomotion. These effects were then attenuated by TRPV1 blockage but not CB1 blockage, indicating some positive involvement of this receptor as well [29]. Activating the ECS may simply amplify its ability to restore balance in an unbalanced state, specifically when that unbalanced states causes a decrease in ECS tone as seen above. The same CB1 polymorphism noted above, which is associated with impulsivity in ADHD, is also associated with some patient's with Tourette's syndrome [30].

12.2.4 ECS and Working Memory

Working memory is another domain of cognition vital to learning and problem-solving. Here, a person is able to temporarily retain a limited amount of information for immediate use. In a human study, the effects on working memory were examined in subjects with a history of cannabis use along with a single nucleotide polymorphism of a CB1 receptor, rs1406977, which results in lower than normal CB1 mRNA in the prefrontal part of the brain. Subjects that had both the SNP-reducing CB1 function and history of cannabis use had worse working memory performance than the other groups [31]. As was seen previously with impairments in procedural learning, polymorphisms in the ATTN repeats of an untranslated section of the CB1 gene also affect working memory, with subjects with fewer repeats having better scores [32]. This indicates a role for the ECS in working memory, a role that is disrupted also by THC intoxication or even chronic cannabis use-related changes in ECS tone.

12.2.5 Acute Versus Chronic Effects of THC on Cognition

The effects of the ECS on cognition are exceeding plastic and both the brain and the ECS are capable of rapid adaptation. In fact, a large systemic review of acute versus chronic effects of cannabis on cognition found that not only did alterations on cognition decrease with repeated exposures but that the development of tolerance in the domain of cognition was highest of all other systems [33]. It is very likely, therefore, that the ECS actually plays a major role in the adaptability of human cognition as well, which is undoubtably one of our species most important cognitive abilities. Our ability to respond to stress with ongoing neuroplasticity demonstrates this ability. Studies have demonstrated that the ventromedial prefrontal cortex (vmPFC) plays a major role in this process. Early studies of stress in rats focused on catecholamines and the hypothalamic-pituitary-adrenal axis, but they were not adequate to fully explain behavioral changes. In the vmPFC, acute stress causes a quick drop in AEA levels before returning to normal 1 h later. 2-AG under the same conditions showed a delayed reaction with increases 1 h post-stress that was mediated by glucocorticoids. Disrupting the ECS, specifically CB1 receptor, in the vmPFC, resulted in abnormal emotional behaviors such as anxiety and reduced behavioral coping responses mediated there [34].

12.3 Neuroprotection

Simply by nature of its typical feedback inhibition function, the endocannabinoid system is ideally suited for neuroprotection. One of the major sources of damage to neurons, particularly in neurodegenerative diseases, is excitotoxicity. This is when the excessive or prolonged release of glutamate results in the dysregulation of calcium homeostasis, leading to oxidative stress, mitochondrial dysfunction, neuroinflammation, degeneration of dendrites, and eventually cell death [35]. Numerous animal studies of excitotoxicity have shown considerable benefit from activation of the ECS. 2-AG was found to be neuroprotective for brain injury in mice with findings of decreased brain edema, reduced infarct volume, and decreased hippocampal cell death, as well as better clinical recovery when administered after injury. Aside from experimental models where exogenous 2-AG was given and the implicit therapeutic potential, it has also been observed that the levels of 2-AG are increased after injury, showing that it is a part of the body's innate mechanism for injury recovery [36]. 2-AG here is not only feeding back to decrease the excessive release of glutamate. It also regulates calcium influx within the cell and decreases inflammation. Both of these functions been seen previously in the chapters on ECS basics and immunology. Thus, 2-AG is addressing the sequelae of the excitotoxicity in addition to stopping it at the source. Excitotoxicity also increases the number of cells expressing CB1 receptor in experimental models, whose activation increased brain growth factor release, increasing neuron cell proliferation in response to injury [37].

12.3.1 The ECS in Alzheimer's Disease

In experimental models of Alzheimer's disease (AD) in mice, it has been noted that while the ECS does not have any effect on the formation amyloid plaques, AEA is capable of reducing its cytotoxic effects. Amyloid plaques trigger the activity of glycogen synthase kinase 3 beta (GSK3 beta), which then phosphorylates tau proteins, which are also part of the pathologic process of AD. AEA reduces GSK3 beta activity in experimental models. AEA also decreases astroglial cells, which are one of the immune cells of the brain, decreasing inflammation by decreasing the release of interferon-gamma [38]. GSK3 beta is likely a key player in AD pathophysiology as its increased activity is linked to declining memory, neurogenesis, and synaptic function in affected subjects [39]. While reductions in CB1 receptors in postmortem cortical brain tissues of AD patients have correlated only with hypophagia, there is an expected increase in CB2 expression [40]. Analysis of tissue of AD patients has found significantly lower levels of AEA but not 2-AG in the mid-frontal and temporal lobes. In fact, there was an inverse relationship between amyloid plaques and AEA, suggesting that the plaques may be impairing AEA mobilization, with further effects on cognition [41]. Given the protective effects of AEA, creating a localized endocannabinoid deficiency would be expected to further disease progression. This deficiency may be mediated by increased expression of FAAH, the enzyme that breaks down AEA, by astroglia cells and microglia cells associated with plaques [42]. CB2 activation on neuro-immune cells decreases their cytotoxicity and improves nerve cell survival as well [43]. In addition to this, activation of CB2 receptor has even been shown to activate macrophages to clear amyloid plaques [44].

12.3.2 ECS in Migraines

Endocannabinoid deficiency in the brain leading to uncontrolled excitation can have other significant pathologic manifestations other than neurodegeneration. Migraines are currently thought to originate via a phenomenon called cortical spreading depression where a self-propagating wave of neuron hyper-excitability followed by transient depression results in aura symptoms such as lights seen in the visual field and then eventually pain and inflammation. Suppression of this phenomenon has been shown clinically with activation of the CB1 receptor in rat brains [45]. Many migraine sufferers routinely use cannabis to address the symptoms of migraine. It has anti-nausea effects seen in the GI chapter. It has antinociceptive effects noted both in this chapter and the chapter covering the PNS and its anti-inflammatory effects noted in the chapter on the immune system. However, it also appears that the ECS plays a role in the prevention of migraine's root cause. Migraines, in fact, were one of the first diseases suspected as being in part due to ECS deficiency. Lower levels of AEA have been found in the CSF of patients with chronic migraines [46]. Migraines as diseases of the endocannabinoid deficiency are further supported by the association between migraine sufferers and CB1R polymorphisms [47]. A review of medical literature also revealed that treatment of migraines with cannabis was effective in reducing the duration and frequency of migraines [48]. Another mechanism for ECS-mediated management of migraines was discussed in the earlier section on centrally mediated pain. CGRP is known to play a key role in pathophysiology of migraine, and blocking CGRP sensitization of the trigeminal nerve can block a migraine attack [49]. AEA prevented the release of CGRP both synaptically and postsynaptically. This was shown to decrease NO release mediated by CGRP, which results in dural blood vessel dilation and headache [50].

12.3.3 ECS in Seizure Disorder

Seizure is another medical condition caused by uncontrolled burst of neuroelectric activity. Given the ECS role in feedback inhibition of glutamate release, it should come as no surprise at this point that dysfunction of the ECS could result in seizures. Eighty percent of mice with both CB1 and CB2 genes knocked out develop epilepsy [51]. Variants in the CB1R and DAGL, the enzyme that breaks down 2-AG, have been shown to produce increased seizures in mice as well [52]. Thus far, the only polymorphism that is related to epilepsy is associated with the gene encoding

FAAH [53]. Interestingly, this polymorphism is not associated with pediatric obesity, indicating a less effective enzyme. It is likely that this polymorphism actually causes an increase in enzyme function, thus decreasing the AEA level [54].

References

1. Prospéro-García O, Amancio-Belmont O, Becerril Meléndez AL, Ruiz-Contreras AE, Méndez-Díaz M. Endocannabinoids and sleep. Neurosci Biobehav Rev. 2016;71:671–9. https://doi.org/10.1016/j.neubiorev.2016.10.005.
2. Martinez-Vargas M, Morales-Gomez J, Gonzalez-Rivera R, Hernandez-Enriquez C, Perez-Arredondo A, Estrada-Rojo F, Navarro L. Does the neuroprotective role of anandamide display diurnal variations? Int J Mol Sci. 2013;14(12):23341–55. https://doi.org/10.3390/ijms141223341.
3. Hanlon EC, Tasali E, Leproult R, Stuhr KL, Doncheck E, de Wit H, Hillard CJ, Van Cauter E. Circadian rhythm of circulating levels of the endocannabinoid 2-arachidonoylglycerol. J Clin Endocrinol Metab. 2015;100(1):220–6. https://doi.org/10.1210/jc.2014-3455.
4. Hanlon EC. Impact of circadian rhythmicity and sleep restriction on circulating endocannabinoid (eCB) N-arachidonoylethanolamine (anandamide). Psychoneuroendocrinology. 2020;111:104471. https://doi.org/10.1016/j.psyneuen.2019.104471.
5. Soria-Gómez E, Guzmán K, Pech-Rueda O, Montes-Rodríguez CJ, Cisneros M, Prospéro-García O. Oleoylethanolamide affects food intake and sleep-waking cycle through a hypothalamic modulation. Pharmacol Res. 2010;61(5):379–84. https://doi.org/10.1016/j.phrs.2010.01.010.
6. Jumpertz R, Wiesner T, Blüher M, Engeli S, Bátkai S, Wirtz H, Bosse-Henck A, Stumvoll M. Circulating endocannabinoids and N-acyl-ethanolamides in patients with sleep apnea – specific role of oleoylethanolamide. Exp Clin Endocrinol Diabetes. 2010;118(9):591–5. https://doi.org/10.1055/s-0030-1253344.
7. Murillo-Rodriguez E, Blanco-Centurion C, Sanchez C, Piomelli D, Shiromani PJ. Anandamide enhances extracellular levels of adenosine and induces sleep: an in vivo microdialysis study. Sleep. 2003;26(8):943–7. https://doi.org/10.1093/sleep/26.8.943.
8. Herrera-Solís A, Vásquez KG, Prospéro-García O. Acute and subchronic administration of anandamide or oleamide increases REM sleep in rats. Pharmacol Biochem Behav. 2010;95(1):106–12. https://doi.org/10.1016/j.pbb.2009.12.014.
9. Pérez-Morales M, De La Herrán-Arita AK, Méndez-Díaz M, Ruiz-Contreras AE, Drucker-Colín R, Prospéro-García O. 2-AG into the lateral hypothalamus increases REM sleep and cFos expression in melanin concentrating hormone neurons in rats. Pharmacol Biochem Behav. 2013;108:1–7. https://doi.org/10.1016/j.pbb.2013.04.006.
10. Gates PJ, Albertella L, Copeland J. The effects of cannabinoid administration on sleep: a systematic review of human studies. Sleep Med Rev. 2014;18(6):477–87. https://doi.org/10.1016/j.smrv.2014.02.005.
11. Schierenbeck T, Riemann D, Berger M, Hornyak M. Effect of illicit recreational drugs upon sleep: cocaine, ecstasy and marijuana. Sleep Med Rev. 2008;12(5):381–9. https://doi.org/10.1016/j.smrv.2007.12.004.
12. Gates P, Albertella L, Copeland J. Cannabis withdrawal and sleep: a systematic review of human studies. Subst Abus. 2016;37(1):255–69. https://doi.org/10.1080/08897077.2015.1023484.
13. Lovell ME, Akhurst J, Padgett C, Garry MI, Matthews A. Cognitive outcomes associated with long-term, regular, recreational cannabis use in adults: a meta-analysis. Exp Clin Psychopharmacol. 2020;28(4):471–94. https://doi.org/10.1037/pha0000326.

14. Broyd SJ, van Hell HH, Beale C, Yücel M, Solowij N. Acute and chronic effects of cannabinoids on human cognition-a systematic review. Biol Psychiatry. 2016;79(7):557–67. https://doi.org/10.1016/j.biopsych.2015.12.002.
15. Herkenham M, Lynn AB, Little MD, Johnson MR, Melvin LS, de Costa BR, Rice KC. Cannabinoid receptor localization in brain. Proc Natl Acad Sci USA. 1990;87(5):1932–6. https://doi.org/10.1073/pnas.87.5.1932.
16. Bossong MG, Jansma JM, van Hell HH, Jager G, Kahn RS, Ramsey NF. Default mode network in the effects of Δ9-Tetrahydrocannabinol (THC) on human executive function. PLoS One. 2013;8(7):e70074. https://doi.org/10.1371/journal.pone.0070074.
17. Vogel A, Wilken-Schmitz A, Hummel R, Lang M, Gurke R, Schreiber Y, Schäfer MKE, Tegeder I. Low brain endocannabinoids associated with persistent non-goal directed nighttime hyperactivity after traumatic brain injury in mice. Sci Rep. 2020;10(1):14929. https://doi.org/10.1038/s41598-020-71879-x.
18. Bossong MG, van Hell HH, Schubart CD, van Saane W, Iseger TA, Jager G, van Osch MJP, Jansma JM, Kahn RS, Boks MP, Ramsey NF. Acute effects of Δ9-tetrahydrocannabinol (THC) on resting state brain function and their modulation by COMT genotype. Eur Neuropsychopharmacol. 2019;29(6):766–76. https://doi.org/10.1016/j.euroneuro.2019.03.010.
19. Uddin LQ. Salience processing and insular cortical function and dysfunction. Nat Rev Neurosci. 2015;16(1):55–61. https://doi.org/10.1038/nrn3857.
20. Grossberg S. Attention: multiple types, brain resonances, psychological functions, and conscious states. J Integr Neurosci. 2021;20(1):197–232. https://doi.org/10.31083/j.jin.2021.01.406.
21. Gangadin SS, Cahn W, Scheewe TW, Hulshoff Pol HE, Bossong MG. Reduced resting state functional connectivity in the hippocampus-midbrain-striatum network of schizophrenia patients. J Psychiatr Res. 2021;138:83–8. https://doi.org/10.1016/j.jpsychires.2021.03.041.
22. Hester R, Nestor L, Garavan H. Impaired error awareness and anterior cingulate cortex hypoactivity in chronic cannabis users. Neuropsychopharmacology. 2009;34(11):2450–8. https://doi.org/10.1038/npp.2009.67.
23. Patros CH, Alderson RM, Kasper LJ, Tarle SJ, Lea SE, Hudec KL. Choice-impulsivity in children and adolescents with attention-deficit/hyperactivity disorder (ADHD): a meta-analytic review. Clin Psychol Rev. 2016;43:162–74. https://doi.org/10.1016/j.cpr.2015.11.001.
24. Ende G, Cackowski S, Van Eijk J, Sack M, Demirakca T, Kleindienst N, Bohus M, Sobanski E, Krause-Utz A, Schmahl C. Impulsivity and aggression in female BPD and ADHD patients: association with ACC glutamate and GABA concentrations. Neuropsychopharmacology. 2016;41(2):410–8. https://doi.org/10.1038/npp.2015.153.
25. Guo B, Xi K, Mao H, Ren K, Xiao H, Hartley ND, Zhang Y, Kang J, Liu Y, Xie Y, Zhou Y, Zhu Y, Zhang X, Fu Z, Chen JF, Hu H, Wang W, Wu S. CB1R dysfunction of inhibitory synapses in the ACC drives chronic social isolation stress-induced social impairments in male mice. Neuron. 2024;112(3):441–57. https://doi.org/10.1016/j.neuron.2023.10.027.
26. Bidwell LC, Metrik J, McGeary J, Palmer RH, Francazio S, Knopik VS. Impulsivity, variation in the cannabinoid receptor (CNR1) and fatty acid amide hydrolase (FAAH) genes, and marijuana-related problems. J Stud Alcohol Drugs. 2013;74(6):867–78. https://doi.org/10.15288/jsad.2013.74.867.
27. Cooper RE, Williams E, Seegobin S, Tye C, Kuntsi J, Asherson P. Cannabinoids in attention-deficit/hyperactivity disorder: a randomised-controlled trial. Eur Neuropsychopharmacol. 2017;27(8):795–808. https://doi.org/10.1016/j.euroneuro.2017.05.005.
28. Jakowiecki J, Abel R, Orzeł U, Pasznik P, Preissner R, Filipek S. Allosteric modulation of the CB1 cannabinoid receptor by Cannabidiol-A molecular modeling study of the N-terminal domain and the allosteric-orthosteric coupling. Molecules. 2021;26(9):2456. https://doi.org/10.3390/molecules26092456.

29. Tzavara ET, Li DL, Moutsimilli L, Bisogno T, Di Marzo V, Phebus LA, Nomikos GG, Giros B. Endocannabinoids activate transient receptor potential vanilloid 1 receptors to reduce hyperdopaminergia-related hyperactivity: therapeutic implications. Biol Psychiatry. 2006;59(6):508–15. https://doi.org/10.1016/j.biopsych.2005.08.019.
30. Szejko N, Fichna JP, Safranow K, Dziuba T, Żekanowski C, Janik P. Association of a variant of CNR1 gene encoding cannabinoid receptor 1 with Gilles de la Tourette syndrome. Front Genet. 2020;11:125. https://doi.org/10.3389/fgene.2020.00125.
31. Colizzi M, Fazio L, Ferranti L, Porcelli A, Masellis R, Marvulli D, Bonvino A, Ursini G, Blasi G, Bertolino A. Functional genetic variation of the cannabinoid receptor 1 and cannabis use interact on prefrontal connectivity and related working memory behavior. Neuropsychopharmacology. 2015;40(3):640–9. https://doi.org/10.1038/npp.2014.213.
32. Ruiz-Contreras AE, Carrillo-Sánchez K, Gómez-López N, Vadillo-Ortega F, Hernández-Morales S, Carnevale-Cantoni A, Espejel-Núñez A, Méndez-Díaz M, Prospéro-García O. Working memory performance in young adults is associated to the AATn polymorphism of the CNR1 gene. Behav Brain Res. 2013;236(1):62–6. https://doi.org/10.1016/j.bbr.2012.08.031.
33. Colizzi M, Bhattacharyya S. Cannabis use and the development of tolerance: a systematic review of human evidence. Neurosci Biobehav Rev. 2018;93:1–25. https://doi.org/10.1016/j.neubiorev.2018.07.014.
34. Worley NB, Hill MN, Christianson JP. Prefrontal endocannabinoids, stress controllability and resilience: a hypothesis. Prog Neuro-Psychopharmacol Biol Psychiatry. 2018;85:180–8. https://doi.org/10.1016/j.pnpbp.2017.04.004.
35. Binvignat O, Olloquequi J. Excitotoxicity as a target against neurodegenerative processes. Curr Pharm Des. 2020;26(12):1251–62. https://doi.org/10.2174/1381612826666200113162641.
36. Panikashvili D, Simeonidou C, Ben-Shabat S, Hanus L, Breuer A, Mechoulam R, Shohami E. An endogenous cannabinoid (2-AG) is neuroprotective after brain injury. Nature. 2001;413(6855):527–31. https://doi.org/10.1038/35097089.
37. Aguado T, Romero E, Monory K, Palazuelos J, Sendtner M, Marsicano G, Lutz B, Guzmán M, Galve-Roperh I. The CB1 cannabinoid receptor mediates excitotoxicity-induced neural progenitor proliferation and neurogenesis. J Biol Chem. 2007;282(33):23892–8. https://doi.org/10.1074/jbc.M700678200.
38. Aso E, Palomer E, Juvés S, Maldonado R, Muñoz FJ, Ferrer I. CB1 agonist ACEA protects neurons and reduces the cognitive impairment of AβPP/PS1 mice. J Alzheimers Dis. 2012;30(2):439–59. https://doi.org/10.3233/JAD-2012-111862.
39. Lauretti E, Dincer O, Praticò D. Glycogen synthase kinase-3 signaling in Alzheimer's disease. Biochim Biophys Acta, Mol Cell Res. 2020;1867(5):118664. https://doi.org/10.1016/j.bbamcr.2020.118664.
40. Solas M, Francis PT, Franco R, Ramirez MJ. CB2 receptor and amyloid pathology in frontal cortex of Alzheimer's disease patients. Neurobiol Aging. 2013;34(3):805–8. https://doi.org/10.1016/j.neurobiolaging.2012.06.005.
41. Jung KM, Astarita G, Yasar S, Vasilevko V, Cribbs DH, Head E, Cotman CW, Piomelli D. An amyloid β42-dependent deficit in anandamide mobilization is associated with cognitive dysfunction in Alzheimer's disease. Neurobiol Aging. 2012;33(8):1522–32. https://doi.org/10.1016/j.neurobiolaging.2011.03.012.
42. Benito C, Núñez E, Tolón RM, Carrier EJ, Rábano A, Hillard CJ, Romero J. Cannabinoid CB2 receptors and fatty acid amide hydrolase are selectively overexpressed in neuritic plaque-associated glia in Alzheimer's disease brains. J Neurosci. 2003;23(35):11136–41. https://doi.org/10.1523/JNEUROSCI.23-35-11136.2003.
43. Cabral GA, Griffin-Thomas L. Emerging role of the cannabinoid receptor CB2 in immune regulation: therapeutic prospects for neuroinflammation. Expert Rev Mol Med. 2009;11:e3. https://doi.org/10.1017/S1462399409000957.
44. Tolón RM, Núñez E, Pazos MR, Benito C, Castillo AI, Martínez-Orgado JA, Romero J. The activation of cannabinoid CB2 receptors stimulates in situ and in vitro beta-amyloid

removal by human macrophages. Brain Res. 2009;1283:148–54. https://doi.org/10.1016/j.brainres.2009.05.098.
45. Kazemi H, Rahgozar M, Speckmann EJ, Gorji A. Effect of cannabinoid receptor activation on spreading depression. Iran J Basic Med Sci. 2012;15(4):926–36.
46. Sarchielli P, Pini LA, Coppola F, Rossi C, Baldi A, Mancini ML, Calabresi P. Endocannabinoids in chronic migraine: CSF findings suggest a system failure. Neuropsychopharmacology. 2007;32(6):1384–90. https://doi.org/10.1038/sj.npp.1301246.
47. Juhasz G, Lazary J, Chase D, Pegg E, Downey D, Toth ZG, Stones K, Platt H, Mekli K, Payton A, Anderson IM, Deakin JF, Bagdy G. Variations in the cannabinoid receptor 1 gene predispose to migraine. Neurosci Lett. 2009;461(2):116–20. https://doi.org/10.1016/j.neulet.2009.06.021.
48. Poudel S, Quinonez J, Choudhari J, Au ZT, Paesani S, Thiess AK, Ruxmohan S, Hosameddin M, Ferrer GF, Michel J. Medical cannabis, headaches, and migraines: a review of the current literature. Cureus. 2021;13(8):e17407. https://doi.org/10.7759/cureus.17407.
49. Iyengar S, Johnson KW, Ossipov MH, Aurora SK. CGRP and the Trigeminal System in Migraine. Headache. 2019;May;59(5):659–681. https://doi.org/10.1111/head.13529. Epub 2019 Apr 14. PMID: 30982963; PMCID: PMC6593989.
50. Akerman S, Kaube H, Goadsby PJ. Anandamide acts as a vasodilator of dural blood vessels in vivo by activating TRPV1 receptors. Br J Pharmacol. 2004;Aug;142(8):1354–60. https://doi.org/10.1038/sj.bjp.0705896. Epub 2004 Jul 26. PMID: 15277315; PMCID: PMC1575202.
51. Rowley S, Sun X, Lima IV, Tavenier A, de Oliveira ACP, Dey SK, Danzer SC. Cannabinoid receptor 1/2 double-knockout mice develop epilepsy. Epilepsia. 2017;58(12):e162–6. https://doi.org/10.1111/epi.13930.
52. Smith DR, Stanley CM, Foss T, Boles RG, McKernan K. Rare genetic variants in the endocannabinoid system genes CNR1 and DAGLA are associated with neurological phenotypes in humans. PLoS One. 2017;12(11):e0187926. https://doi.org/10.1371/journal.pone.0187926.
53. Anvar LH, Alejafar A, Moosavi SE, Charsouei S, Zeynalzadeh N, Fanid LM, Emamalizadeh B, Aydinlou ZH, Vaezi H, Kashefi A, Tomaz C, Nikanfar M, Ahmadalipour A. The study of rs324420 (C385A) polymorphism of the FAAH gene of the endocannabinoid system in patients with epilepsy and ADHD. Epilepsy Res. 2023 May;192:107100. https://doi.org/10.1016/j.eplepsyres.2023.107100. Epub 2023 Feb 3. PMID: 37018974.
54. Müller TD, Brönner G, Wandolski M, Carrie J, Nguyen TT, Greene BH, Scherag A, Grallert H, Vogel CI, Scherag S, Rief W, Wichmann HE, Illig T, Schäfer H, Hebebrand J, Hinney A. Mutation screen and association studies for the fatty acid amide hydrolase (FAAH) gene and early onset and adult obesity. BMC Med Genet. 2010;11:2. https://doi.org/10.1186/1471-2350-11-2.

Mood, Motivation, and Mental Health 13

Given the number and concentration of cannabinoid receptor seen in the brain, it would be rather remarkable if they were not also involved in the pathophysiology of mental illnesses. Cannabis has been used for millennia to treat such conditions and remains one of the most common forms of self-medication among depression, anxiety, and addiction sufferers. Their role in mood disorders cannot be understated and as we will see, without them, we would all be intensely miserable beings. We will begin this chapter with discussion first of major depression disorder and then tackle anxiety and finally addiction, psychosis, and schizophrenia.

13.1 ECS in Major Depression Disorder

Major depression disorder (MDD) is diagnosed clinically by criteria outlined in the DMS V, with patients being found to have at least five of nine different symptoms including anhedonia or depressed mood, weight loss, increased or decreased appetite, trouble sleeping or excessive sleepiness, psychomotor retardation or agitation, fatigue or loss of energy, feelings of worthlessness or guilt, trouble concentrating, and finally recurrent thoughts of death or suicide. Symptoms must be present for at least 2 weeks. Given the variety of possible symptoms, the variability in presentation of MDD is also therefore extremely diverse and affects many different parts of the brain. MDD's sequelae extend beyond the brain and mood to changes in the immune, endocrine, and cardiovascular systems. Modern pharmacotherapy treatment of depression is based on the monoamine deficiency theory. This attributed the cause of depression to deficiencies of monoamine neurotransmitters in the brain, first catecholamines like norepinephrine and then serotonin, and treated it with drugs that could increase the levels of these neurotransmitters in the brain. While this focus has resulted in many drugs that are partially successful in treating depression symptoms, they are by no means a cure, and recent evidence has shown MDD to be a multifactorial condition with many different genetic and environmental risks

J.-P. H. Dedam et al., *Diseases of the Endocannabinoid System*,
https://doi.org/10.1007/978-3-031-87484-0_13

such as early exposure to trauma that expand beyond just serotonin deficiency [1]. As we will see below, the endocannabinoid system plays an important role in nearly all identified risks and sequelae, tying it to the pathophysiology of mood disorders perhaps more than any of the other disease processes we have reviewed thus far.

There is ample evidence in animals for a role of the ECS in depression with CB1 knockout mice developing anhedonia and impaired emotional reactions to stress [2]. In other studies in rats, low doses of CB1 agonist were shown to have antidepressant effects. These effects were accompanied by increases in serotonergic nerve activity in the dorsal raphe nucleus. This is the largest serotoninergic nucleus in the brain and responsible for a large proportion of the serotonin transmission of the forebrain. The above antidepressant effect of CB1 agonist was blocked both by the CB1 blocker rimonabant and by serotonin depletion of the brain. High doses of CB1 activation actually decreased serotonin output independent of CB1 [3], indicating a biphasic effect similar to the known biphasic effect of cannabinoids on anxiety that we will discuss later. In a separate experiment, increased AEA levels produced by experimentally blocking FAAH not only increase serotonin from the dorsal raphe nucleus but also increased noradrenergic firing in the locus coeruleus [4]. These findings in mice and rats are consistent with human trials of rimonabant, which was studied as a weight loss drug. Given the metabolic effects of the ECS seen in Chap. 4, it is no surprise that it was an effective treatment for weight loss. However, the adverse psychiatric side effects prevented it from ever reaching the market as it increased risk of depression, anxiety, and suicide [5]. Subsequent research on rimonabant in rats has shown that chronic administration results in decreased frontal cortex serotonin levels as well [6]. While serotonin levels in the brain probably do not completely explain the pathophysiology of depression, they are important for normal mood, and they are very much dependent on a functioning endocannabinoid system.

Studies in rats looking at different antidepression treatment, e.g., repetitive transcranial magnetic stimulation (rTMS), showed that it resulted in higher CB1 expression in the brain, with increased cell proliferation and increased brain-derived neurotrophic factor, an important nerve growth factor for long-term memory. The beneficial effects of rTMS were blocked by CB1 receptor blockers [7]. Subsequent studies by the same researcher also confirmed an increase in 2-AG in the hippocampus that correlated with the noted decrease in depression symptoms [8]. Depressive symptoms in both studies were induced by chronic unpredictable mild stress. It definitely appears that the ECS plays a very important role in dealing this type of stress.

Low serum levels of 2-AG have also been noted in the blood of women with depression, and lower levels were correlated with longer duration of the episode [9]. In a separate study in humans with depression treated with electroconvulsive treatment, lower levels of ECS correlated with depressive, anxiety, and suicidal symptoms. At baseline, low endocannabinoid levels predicted worse symptoms, and these levels increased after six ECT treatments along with improvement in symptoms [10]. Perhaps the most compelling evidence though comes from the study of polymorphisms identified in the endocannabinoid system and their relation to depression. There is already evidence that the risk of depression can be influenced by genetics [11]. In one study looking at polymorphisms of CB1 receptor gene in patients with MDD, subjects with the rs806371 polymorphism were more likely to have melancholia and psychotic symptoms and a higher risk of not obtaining remission with treatment. This SNP is located in the promoter region of the gene, and a previous study demonstrated that it results in lower expression of CB1 receptor [12]. This same SNP was also associated with higher rates of depression following trauma in a recent study in humans [13]. Another polymorphism of CB1 receptor predicted greater depression symptoms in response to life events [14]. The relationship between the ECS and serotonin in depression pathogenesis is also strengthened by evidence that a polymorphism of CB1 receptor also confers greater risk of SSRI treatment failure, suggesting again that CB1 receptor is necessary in humans for proper serotonin release, likely through mechanisms discussed earlier in the chapter [15]. Conversely, the rs2023239 polymorphism of the CB1 receptor gene, which has been shown to increase CB1 density on blood lymphocytes [16], was shown to be protective of MDD among methadone users [17].

The association between polymorphisms of the ECS and depression is not limited to just CB1 receptor. The CB2 receptor polymorphism R63Q was associated with increased anxious and depressive phenotype risk in children exposed to trauma [18]. Studies in mice have also revealed that activation of CB2 by beta-caryophyllene also had an anxiolytic and antidepressant effect [19]. Sustained activation of CB2 receptor has also been shown to upregulate serotonin receptors [20]. However, as mentioned previously, depression is more than just a simple serotonin deficiency syndrome. Neuroinflammation is now beginning to be understood as playing an important role in the pathogenesis of depression. Depressive symptoms are indeed very common in most neuro-inflammatory/neurodegenerative diseases [21]. Meta-analysis of patients with MDD showed elevations in IL-6 and CRP, two markers of inflammation [22]. And activation of CB2 receptor by AEA has been shown to reduce neuroinflammation in the brain, specifically in this study as measured by IL-6 and NO production [23]. As seen in numerous other systems, the ECS, specifically activation of CB2 receptor is a vital part of the body's response to inflammation, and decreases in ECS tone result in increased inflammation and in this instance, greater depression symptomatology.

Disruption of the hypothalamic pituitary adrenal axis is an important feature of MDD. Early life stress is a significant risk factor for the development of MDD. The HPA axis can become sensitized by repeated insults, resulting in overactivation of the axis, leading to excessive cortisol release and the myriad of consequences that this overproduction results in. These same disruptions in the HPA axis are also seen in chronic depression, and persistent HPA axis disruption is associated with higher rates of relapse and chronicity of symptoms [24]. The ECS is a key regulator of the HPA axis, as evidenced by significant HPA hyperactivity in CB1 knockout mice [25]. This is also supported by the human polymorphism studies associating CB1 receptor with risk of depression following trauma and other life stressors as mentioned above. CB1 activation in the prefrontal cortex is necessary for termination of the stress response. Cortisol leads to accumulation of 2-AG here, which then inhibits GABA transmission, likely then increasing outflow to the prelimbic region to terminate stress response by deactivating the HPA axis [26]. However, chronic stress and overactivation of the HPA also affect the ECS and can disrupt the normal feedback inhibition loop. Corticotropin-releasing hormone, produced in the hypothalamus and is the first signal in the cascade to produce more cortisol, increases FAAH activity in the amygdala, thus reducing AEA levels and increasing anxiety in mice [27].

Finally, another area of study in the pathogenesis of MDD is hippocampal atrophy in untreated MDD patients [28]. This can be reversed by fluoxetine in animal models, leading to cell proliferation mediated by BDNF [29, 30].

CB1 activation stimulates hippocampus neurogenesis in mice [31]. While the exact mechanism is not known, it may be related to BDNF, which is reduced in CB1 knockout mice, leading to enhanced despair behavior [32].

As seen above, the ECS is involved in the many pathogenic pathways of MDD, almost exclusively preventing them. Given the significant association between many polymorphisms and MDD, as well as the subjective antidepressant effects experienced by cannabis users, there can be no doubt that a significant proportion of MDD is due to endocannabinoid deficiency. More research needs to be done in how treatment of the ECS can be integrated into the overall treatment of MDD.

13.2 ECS and Anxiety

Anxiety symptoms were mentioned frequently in the section above, as they are often seen along with depression. The American Psychology Association defines anxiety as an emotion characterized by feelings of tension and worried thoughts in response to a diffuse, undefined threat at some point in the future. Experimental models have long shown a biphasic effect of the ECS on anxiety symptoms. With lower amounts of CB1 activation, anxiety symptoms are decreased, while higher doses sometimes increased it. There does appear to be a tonic effect of the ECS on anxiety, as CB1 receptor mice have increased anxiety at baseline [33]. Mice without the gene for DAGL, the enzyme that produces 2-AG, also had increased anxiety [34]. The areas of the brain that control anxiety symptoms, i.e., the amygdala, the hippocampus, and prefrontal cortex, all express large numbers of CB1 receptors as we have already reviewed in this chapter and the previous one. They are found on both glutamatergic and GABAergic neurons. Generally speaking, functioning as a feedback inhibitor, ECS activation on glutamatergic neurons decreases anxiety symptoms by inhibiting anxiogenic glutamatergic nerve transmission. Activation of the ECS on GABAergic neurons causes anxiety by inhibiting anxiolytic GABAergic nerve transmission. GABAergic neurons are actually more numerous and more sensitive to CB1 activation than glutamatergic receptors. This on its own should favor lower doses of CB1 activation increasing anxiety, as more GABAergic neurons would be inhibited first but that not what is seen in vivo. A possible explanation for this is that GABAergic neurons are actually being tonically inhibited already by CB1, so a small increase in CB1 activation makes a little difference, whereas glutamatergic neurons are not tonically inhibited, so the addition of CB1 activation results in a larger change in their overall inhibition. And, in this case, a greater decrease in glutamatergic than GABAergic firing results in anxiolysis. In another study in mice, low-dose cannabis analytic effects were confirmed by the finding that his effect was not seen in mice with CB1 receptor deactivated on glutamate neurons, while high-dose cannabinoid angiogenic effects were not seen in mice with CB1 receptor inactivated on glutamate neurons [35]. To complicate matters further, the ECS has been shown to produce anxiolysis that requires the activation of GABA receptors in the forebrain of mice to produce the effect. These would mean that there are definitely other pathways and interactions in the brain between CB1 and GABA that result in anxiolysis that have yet to be elucidated [36].

Another explanation for the biphasic effect of the ECS on anxiety is the colocation of TRPV1 receptors with CB1 receptors in the periaqueductal gray matter. TRPV1 receptors have an opposing intracellular effect on calcium to CB1 receptor: increasing it, while the latter decreasing it. CB1 is expressed in much higher quantities, so at low doses of their common ligand, AEA, CB1 activation dominates and the result is anxiolysis. At higher concentrations, TRPV1 dominates and panic ensues. This panic response can be blocked by TRPV1 antagonists [37]. From an evolutionary standpoint, this makes a certain sense. Under low stress conditions, its best to remain calm, but once a situation reaches a certain threshold of danger and stress, the best approach is to get hell out of there. TRPV1 knockout mice have also been shown to have increased resiliency to chronic unpredictable stress, a key risk factor for the development of anxiety disorders. This would indicate that TRPV1 activation plays a role in the development of chronic anxiety conditions in addition to acute anxiety symptoms [38]. Panic disorder in human has also been linked to polymorphisms of CB1 gene, specifically rs1049353 and rs806368 polymorphisms [39]. In a separate study linking rs1049353 with depression risk, the researchers also found lower 2-AG levels, indicating an underperforming ECS [40].

Animal models have shown that the functional coupling of circuits between the amygdala and the prefrontal cortex, specifically the dorsal anterior cingulate gyrus, can result in anxiety. Studies in mice showed that repeat stress and trauma increase glutamatergic signaling between the two and that 2-AG functioned as the main inhibitor of this pathway. Under repeated traumatic events, however, 2-AG levels dropped, resulting in increased anxiety behavior [41]. A similar connection also exists between the amygdala and the hippocampus. In this experimental model, chronic stress increased levels of an epigenetic transcriptional factor lysine-specific demethylase-1, which increased expression of MAGL, the enzyme that degrades 2-AG. This resulted in the same increase in glutamatergic signaling seen between the amygdala and the prefrontal cortex and loss of ECS control over excessive angiogenic signaling [42]. Another study using the human polymorphism in the FAAH gene, which results in reduced expression and higher AEA levels, was genetically moved into experimental mice. This human SNP in mice served to characterize a different prefrontal cortex and amygdala connection. This connection was between the ventral medial PFC and the amygdala. The study showed that this connection had the opposite effect with increased AEA resulting in greater connectivity of the amygdala and the VMPFC and then resulting in decreased anxiety [43].

Serotonin also plays an important role in anxiolysis, and as was noted numerous times in the previous section on depression, the ECS is necessary for the proper functioning of serotonin within the brain. The medial prefrontal cortex has a direct synaptic connection to the dorsal raphe nucleus serotonin and GABA neurons. Activation of these serotoninergic neurons causes serotonin release all over the brain. CB1 receptors are present presynaptically for both neurons, but the GABA neurons, which will inhibit serotonin release in the brain, are more sensitive to this input than the serotonergic neurons. Activation of CB1 receptors here therefore favors excitation of the serotonin neurons by removing its GABAergic inhibition [44]. A study of human polymorphism of the CB1 promoter region that results in lower CB1 expression was found to be associated with anxiety symptoms in homozygous patients and, when found in conjunction with a polymorphism of the 5HT serotonin receptor promoter region, resulted in either very high or very low serotonin levels with worsen anxiety in either cases [45]. Deletion of FAAH gene in mice, which would increase the available AEA, also improves anxiety along with increasing serotonin levels in the frontal cortex [46]. This finding definitely links the ECS and serotoninergic neurons in their role in controlling anxiety.

Aside from the mostly anxiolytic effects discussed above, CB1 receptor is also necessary to transition from reactive, phasic fear responses to sustained anxious states that characterize anxiety disorders. Presynaptic CB1 receptors on amygdala projections to the bed nucleus of the stria terminalis are necessary to make this transition under conditions of unpredictable stressors [47]. Anxiety disorders are more common among cannabis users and likely this is mostly preselection by patients who are already living with anxiety and self-medicating, but the above mechanism could represent a pathway by which cannabis could possibly potentiate the development of an anxiety disorder by facilitating transition from phasic fear to more sustained anxious states.

CB2 activation also appears to have anxiolytic effects. Blockade of CB2 receptors in mice increased anxious behavior [48]. CB2 knockout mice were noted to have lower serotonin levels in the dorsal raphe nucleus, anxiety, and other symptoms similar to schizophrenia [49]. A human polymorphism in CB2 gene that decreases its function were also associated with anxiety, as noted in the previous section on depression [18]. As with depression, there can really be no doubt that disruption of the ECS can result in anxiety. Some anxiety conditions may be due to a simple endocannabinoid deficiency, but given the complexity of the regulation of anxiety by the ECS, many more anxiety conditions are the result of more mixed or localized endocannabinoid system dysfunction.

13.3 ECS and Addiction

Perhaps one of the best studied areas of action of cannabis on the brain is the reward system, predominately due to the fact that for the last 50 years, the only US agency funding studies on it has been the National Institute on Drug Abuse. And their agenda has been to show what a dangerous and addictive drug it is. The reward system begins with the ventral tegmental area (VTA), a collection of neurons in the midbrain, and A10 dopaminergic neurons that project to the nucleus accumbens, a part of the basal ganglia. Both regions contain CB1 receptors. These neurons release dopamine in response to reward, be it food, sex, or drugs. These neurons have two firing patterns: tonic firing, which determines motivation to the reward, and phasic bursts that establish long-term memories of the reward. CB1 receptor activation in the VTA increases both tonic and phasic firing, despite the dopaminergic neurons not actually having receptors. Instead, CB1 receptors are thought to inhibit inhibitory GABAergic interneurons, thus allowing more dopamine release. Increased firing of these neurons leads to intracellular production of 2-AG, which leads to more firing. As was seen in the chapter on obesity, when ECS is involved in feedforward pathways rather than the typical feedback inhibition, problems can arise. Within the nucleus accumbens, prefrontal glutamatergic neurons project onto D1 medium spiny neurons, prompting them to produce 2-AG, which feedbacks to inhibit glutamate release. Long-term depression of these prefrontal inputs facilitates drug-seeking behavior [50]. CB1 receptor activity may be balanced by the presence of CB2 receptor in VTA dopamine neurons. Activation of these receptors in the nucleus accumbens decreased dopamine release in response to cocaine in mouse studies [51]. A polymorphism of CB2 gene in humans known to reduce function has been linked to alcoholism [52]. Cannabis, unsurprisingly, increases dopamine release in the nucleus accumbens, prompting drug-seeking behavior, which is reduced by CB1 receptor antagonists like rimonabant [53]. CBD, as an allosteric inhibitor of CB1, has also been shown in animal models to decrease dopamine release by methamphetamines, a much more potent potentiator of dopamine release than cannabis and therefore a much more potent drug of abuse [54]. Whether this potentially anti-addictive property of CBD functions in as well humans has yet to be elucidated, but it is not uncommon for cannabis users to substitute high CBD strains for high THC strains when trying to cut back on their consumption. While cannabis does not have anywhere near the addiction potential of nicotine, heroin, or cocaine, as seen above, it does have addiction potential. Higher potency strains and synthetic cannabinoids with significantly higher CB1 bonding affinities would only increase that potential. Polymorphisms within the ECS have also been linked to addiction. The Rs2023239 polymorphism in CB1 gene, which increases receptor density and overall ECS tone, has been linked to increased risk of alcohol dependence [55] and nicotine [56]. This raises the possibility that states of endocannabinoid excess could lead to increased risk of dependence.

13.4 ECS and Psychosis

Another psychiatric condition linked to increased endocannabinoid and dopaminergic tone is psychosis. Psychosis can be caused by a myriad of different causes, cannabis intoxication being a very well-documented one. Alterations in endocannabinoid function during acute psychosis have been noted, specifically increased CB1 activation and decreased CB2 activity as well. Increased levels of dopamine are associated with psychosis, and CB1 activation increases this in the mesolimbic areas of the brain, while CB2 decreases it [57]. As was noted in the section on cognition, activation of CB1 receptors in the insular cortex can disrupt salience processing, which helps our brains decide which stimuli to focus on and is another area of the brain affected during psychotic episodes. A polymorphism in the gene for FAAH, which appears to decrease its function and increase AEA, was associated with higher risk of early psychosis but only in subjects with history of cannabis use [58]. This represents an instance of vulnerability due to increased endocannabinoid tone that only becomes clinically relevant with the exposure to even more exogenous cannabinoids. A polymorphism of a CB1 gene, RS 806379, which causes increased risk of alcohol dependence and presumably higher ECS tone, was also shown in that study to be a risk factor for psychosis after methamphetamine use, a state of dramatically increased dopamine in the brain as well [59]. Excessive cannabis use in and of itself has been known to clinically cause a transient psychotic episode that passes once the intoxication passes, but probably of greater clinical significance is the potentiation of early-onset psychosis in patients who are already at risk of developing it on their own [60]. Large meta-analysis has also shown that increasing doses of cannabis use are associated with greater risk of psychosis [61].

13.5 ECS and Schizophrenia

There is also significant evidence of dysregulation of the ECS in schizophrenia. As noted above, cannabis use is associated with increased risk of psychosis. AEA levels were also found to be significantly elevated in the CSF of schizophrenic patients with their first episode of psychosis. Higher levels of AEA were actually protective against psychosis, though, indicating a potential compensatory mechanism in the body rather than a causal effect [62]. CB1 and CB2 receptors are present in the excitatory and inhibitory glutamatergic and GABAergic neurons in the striatum. High levels of dopaminergic activity here are thought to be the cause of psychosis in schizophrenia. Cannabis use was found to decrease dopamine synthesis in the striatum of chronic users who had experienced psychosis, indicating a potential feedback inhibition role. Given the contradictory findings in cannabis-induced psychosis and psychosis in schizophrenia, the specific neurotransmitter concentrations that result psychosis in these two conditions may not be the same [63]. As CB1 receptors are present on both inhibitory and excitatory neurons controlling the dopaminergic neurons of the striatum, much more study needs to be done to better understand the role of ECS dysfunction in schizophrenia. High cannabis use among schizophrenics was found to significantly decrease AEA levels in the CSF but did not significantly alter AEA levels of healthy volunteers [64]. Perhaps schizophrenics are drawn to cannabis as it may help initially relieve some symptoms related to dopamine imbalance but in the long term may actually have negative effects by impairing their bodies' own response to excess dopamine. It may also be possible that an impaired ECS response is part of their underlying risk for schizophrenia and they are merely trying to compensate. Expression of CB1 receptor is lower in the prefrontal cortex of schizophrenics [65]. No CB1 polymorphisms have been identified that confer a higher risk for schizophrenia, but one 135G/A did confer a reduced response to antipsychotic medications [66]. The connection between schizophrenia and cannabis may be partly circumstantial, as substance use rates among schizophrenics are extremely high (greater than 40%) and cannabis, being one of the more readily available drugs, is simply present and easily available. Rates of cannabis use are the same as other illicit drugs in this population, 26 versus 27%, indicating known predisposition to addictive substances known to be present in schizophrenics [67]. As was seen in the previous chapter on cognition, it is possible that the excessive ECS tone seen in CSF of schizophrenics may exacerbate or cause some of the negative systems seen in the disease as well, including poor executive function, attention, and working memory. In fact, one study showed that increased CB1 binding in the striatum of schizophrenics was related with worsened negative symptoms [68]. We have thus far focused on CB1 receptor in schizophrenia; however, new evidence has also implicated CB2 receptor in the disease pathology with CB2 knockout mice displaying many signs of the disease such as impaired memory, hyperlocomotion anxiety, and depression [69]. While a great deal of research still needs to be done, it is clear that the ECS is dysfunctional in schizophrenia. How to address that dysfunction would definitely require a much better understanding of the ECS's role in the disease itself.

References

1. Gallego-Landin I, García-Baos A, Castro-Zavala A, Valverde O. Reviewing the Role of the Endocannabinoid System in the Pathophysiology of Depression. Front Pharmacol. 2021;12:762738. https://doi.org/10.3389/fphar.2021.762738.
2. Martin M, Ledent C, Parmentier M, Maldonado R, Valverde O. Involvement of CB1 cannabinoid receptors in emotional behaviour. Psychopharmacology. 2002;159(4):379–87. https://doi.org/10.1007/s00213-001-0946-5.
3. Bambico FR, Katz N, Debonnel G, Gobbi G. Cannabinoids elicit antidepressant-like behavior and activate serotonergic neurons through the medial prefrontal cortex. J Neurosci. 2007;27(43):11700–11. https://doi.org/10.1523/JNEUROSCI.1636-07.2007.
4. Gobbi G, Bambico FR, Mangieri R, Bortolato M, Campolongo P, Solinas M, Cassano T, Morgese MG, Debonnel G, Duranti A, Tontini A, Tarzia G, Mor M, Trezza V, Goldberg SR, Cuomo V, Piomelli D. Antidepressant-like activity and modulation of brain monoaminergic transmission by blockade of anandamide hydrolysis. Proc Natl Acad Sci USA. 2005;102(51):18620–5. https://doi.org/10.1073/pnas.0509591102.
5. Christensen R, Kristensen PK, Bartels EM, Bliddal H, Astrup A. Efficacy and safety of the weight-loss drug rimonabant: a meta-analysis of randomised trials. Lancet. 2007;370(9600):1706–13.
6. Beyer CE, Dwyer JM, Piesla MJ, Platt BJ, Shen R, Rahman Z, Chan K, Manners MT, Samad TA, Kennedy JD, Bingham B, Whiteside GT. Depression-like phenotype following chronic CB1 receptor antagonism. Neurobiol Dis. 2010;39(2):148–55. https://doi.org/10.1016/j.nbd.2010.03.020.
7. Wang HN, Wang L, Zhang RG, Chen YC, Liu L, Gao F, Nie H, Hou WG, Peng ZW, Tan Q. Anti-depressive mechanism of repetitive transcranial magnetic stimulation in rat: the role of the endocannabinoid system. J Psychiatr Res. 2014;51:79–87. https://doi.org/10.1016/j.jpsychires.2014.01.004.

8. Fang G, Wang Y. Effects of rTMS on hippocampal endocannabinoids and depressive-like behaviors in adolescent rats. Neurochem Res. 2018;43(9):1756–65. https://doi.org/10.1007/s11064-018-2591-y.
9. Hill MN, Miller GE, Ho WS, Gorzalka BB, Hillard CJ. Serum endocannabinoid content is altered in females with depressive disorders: a preliminary report. Pharmacopsychiatry. 2008;41(2):48–53. https://doi.org/10.1055/s-2007-993211.
10. Bloemhof-Bris E, Meiri D, Sulimani L, Genesh SN, Wexler G, Cohen I, Salama A, Burshtein I, Hirschmann S, Feffer K, Weizman S, Stryjer R, Shelef A. Alterations in plasma endocannabinoid concentrations among individuals with major depression treated with electroconvulsive therapy. Psychiatry Res. 2024;337:115967. https://doi.org/10.1016/j.psychres.2024.115967.
11. Mondimore FM, Zandi PP, Mackinnon DF, McInnis MG, Miller EB, Crowe RP, Scheftner WA, Marta DH, Weissman MM, Levinson DF, Murphy-Ebenez KP, Depaulo JR, Potash JB. Familial aggregation of illness chronicity in recurrent, early-onset major depression pedigrees. Am J Psychiatry. 2006;163(9):1554–60. https://doi.org/10.1176/ajp.2006.163.9.1554.
12. Feng Q, Vickers KC, Anderson MP, Levin MG, Chen W, Harrison DG, Wilke RA. A common functional promoter variant links CNR1 gene expression to HDL cholesterol level. Nat Commun. 2013;4:1973. https://doi.org/10.1038/ncomms2973.
13. Fitzgerald JM, Chesney SA, Lee TS, Brasel K, Larson CL, Hillard CJ, deRoon-Cassini TA. Circulating endocannabinoids and prospective risk for depression in trauma-injury survivors. Neurobiol Stress. 2021;14:100304. https://doi.org/10.1016/j.ynstr.2021.100304.
14. Juhasz G, Chase D, Pegg E, Downey D, Toth ZG, Stones K, Platt H, Mekli K, Payton A, Elliott R, Anderson IM, Deakin JF. CNR1 gene is associated with high neuroticism and low agreeableness and interacts with recent negative life events to predict current depressive symptoms. Neuropsychopharmacology. 2009;34(8):2019–27. https://doi.org/10.1038/npp.2009.19.
15. Domschke K, Dannlowski U, Ohrmann P, Lawford B, Bauer J, Kugel H, Heindel W, Young R, Morris P, Arolt V, Deckert J, Suslow T, Baune BT. Cannabinoid receptor 1 (CNR1) gene: impact on antidepressant treatment response and emotion processing in major depression. Eur Neuropsychopharmacol. 2008;18(10):751–9. https://doi.org/10.1016/j.euroneuro.2008.05.003.
16. Ketcherside A, Noble LJ, McIntyre CK, Filbey FM. Cannabinoid receptor 1 gene by cannabis use interaction on CB1 receptor density. Cannabis Cannabinoid Res. 2017;2(1):202–9. https://doi.org/10.1089/can.2017.0007.
17. Icick R, Peoc'h K, Karsinti E, Ksouda K, Hajj A, Bloch V, Prince N, Mouly S, Bellivier F, Lépine JP, Laplanche JL, Vorspan F. A cannabinoid receptor 1 polymorphism is protective against major depressive disorder in methadone-maintained outpatients. Am J Addict. 2015;24(7):613–20. https://doi.org/10.1111/ajad.12273.
18. Lazary J, Eszlari N, Juhasz G, Bagdy G. A functional variant of CB2 receptor gene interacts with childhood trauma and FAAH gene on anxious and depressive phenotypes. J Affect Disord. 2019;257:716–22. https://doi.org/10.1016/j.jad.2019.07.083.
19. Bahi A, Al Mansouri S, Al Memari E, Al Ameri M, Nurulain SM, Ojha S. β-Caryophyllene, a CB2 receptor agonist produces multiple behavioral changes relevant to anxiety and depression in mice. Physiol Behav. 2014;135:119–24. https://doi.org/10.1016/j.physbeh.2014.06.003.
20. Franklin JM, Carrasco GA. G-protein receptor kinase 5 regulates the cannabinoid receptor 2-induced up-regulation of serotonin 2A receptors. J Biol Chem. 2013;288(22):15712–24. https://doi.org/10.1074/jbc.M113.454843.
21. Troubat R, Barone P, Leman S, Desmidt T, Cressant A, Atanasova B, Brizard B, El Hage W, Surget A, Belzung C, Camus V. Neuroinflammation and depression: a review. Eur J Neurosci. 2021;53(1):151–71. https://doi.org/10.1111/ejn.14720.
22. Haapakoski R, Mathieu J, Ebmeier KP, Alenius H, Kivimäki M. Cumulative meta-analysis of interleukins 6 and 1β, tumour necrosis factor α and C-reactive protein in patients with major depressive disorder. Brain Behav Immun. 2015;49:206–15. https://doi.org/10.1016/j.bbi.2015.06.001.
23. Malek N, Popiolek-Barczyk K, Mika J, Przewlocka B, Starowicz K. Anandamide, acting via CB2 receptors, alleviates LPS-induced neuroinflammation in rat primary microglial cultures. Neural Plast. 2015;2015:130639. https://doi.org/10.1155/2015/130639.

24. Juruena MF. Early-life stress and HPA axis trigger recurrent adulthood depression. Epilepsy Behav. 2014;Sep;38:148–59. https://doi.org/10.1016/j.yebeh.2013.10.020. Epub 2013 Nov 20. PMID: 24269030.
25. Barna I, Zelena D, Arszovszki AC, Ledent C. The role of endogenous cannabinoids in the hypothalamo-pituitary-adrenal axis regulation: in vivo and in vitro studies in CB1 receptor knockout mice. Life Sci. 2004;75(24):2959–70. https://doi.org/10.1016/j.lfs.2004.06.006.
26. Hill MN, McLaughlin RJ, Pan B, Fitzgerald ML, Roberts CJ, Lee TT, Karatsoreos IN, Mackie K, Viau V, Pickel VM, McEwen BS, Liu QS, Gorzalka BB, Hillard CJ. Recruitment of prefrontal cortical endocannabinoid signaling by glucocorticoids contributes to termination of the stress response. J Neurosci. 2011;31(29):10506–15. https://doi.org/10.1523/JNEUROSCI.0496-11.2011.
27. Gray JM, Vecchiarelli HA, Morena M, Lee TT, Hermanson DJ, Kim AB, McLaughlin RJ, Hassan KI, Kühne C, Wotjak CT, Deussing JM, Patel S, Hill MN. Corticotropin-releasing hormone drives anandamide hydrolysis in the amygdala to promote anxiety. J Neurosci. 2015;35(9):3879–92. https://doi.org/10.1523/JNEUROSCI.2737-14.2015.
28. Sheline YI, Wang PW, Gado MH, Csernansky JG, Vannier MW. Hippocampal atrophy in recurrent major depression. Proc Natl Acad Sci USA. 1996;93(9):3908–13. https://doi.org/10.1073/pnas.93.9.3908.
29. Malberg JE, Duman RS. Cell proliferation in adult hippocampus is decreased by inescapable stress: reversal by fluoxetine treatment. Neuropsychopharmacology. 2003;28(9):1562–71. https://doi.org/10.1038/sj.npp.1300234.
30. Santarelli L, Saxe M, Gross C, Surget A, Battaglia F, Dulawa S, Weisstaub N, Lee J, Duman R, Arancio O, Belzung C, Hen R. Requirement of hippocampal neurogenesis for the behavioral effects of antidepressants. Science. 2003;301(5634):805–9. https://doi.org/10.1126/science.1083328.
31. Andres-Mach M, Haratym-Maj A, Zagaja M, Rola R, Maj M, Chrościńska-Krawczyk M, Luszczki JJ. ACEA (a highly selective cannabinoid CB1 receptor agonist) stimulates hippocampal neurogenesis in mice treated with antiepileptic drugs. Brain Res. 2015;1624:86–94. https://doi.org/10.1016/j.brainres.2015.07.028.
32. Aso E, Ozaita A, Valdizán EM, Ledent C, Pazos A, Maldonado R, Valverde O. BDNF impairment in the hippocampus is related to enhanced despair behavior in CB1 knockout mice. J Neurochem. 2008;105(2):565–72. https://doi.org/10.1111/j.1471-4159.2007.05149.x.
33. Haller J, Bakos N, Szirmay M, Ledent C, Freund TF. The effects of genetic and pharmacological blockade of the CB1 cannabinoid receptor on anxiety. Eur J Neurosci. 2002;16(7):1395–8. https://doi.org/10.1046/j.1460-9568.2002.02192.x.
34. Shonesy BC, Bluett RJ, Ramikie TS, Báldi R, Hermanson DJ, Kingsley PJ, Marnett LJ, Winder DG, Colbran RJ, Patel S. Genetic disruption of 2-arachidonoylglycerol synthesis reveals a key role for endocannabinoid signaling in anxiety modulation. Cell Rep. 2014;9(5):1644–53. https://doi.org/10.1016/j.celrep.2014.11.001.
35. Rey AA, Purrio M, Viveros MP, Lutz B. Biphasic effects of cannabinoids in anxiety responses: CB1 and GABA(B) receptors in the balance of GABAergic and glutamatergic neurotransmission. Neuropsychopharmacology. 2012;37(12):2624–34. https://doi.org/10.1038/npp.2012.123.
36. Fuss J, Steinle J, Bindila L, Auer MK, Kirchherr H, Lutz B, Gass P. A runner's high depends on cannabinoid receptors in mice. Proc Natl Acad Sci U S A. 2015;112(42):13105–8. https://doi.org/10.1073/pnas.1514996112. Epub 2015 Oct 5. PMID: 26438875; PMCID: PMC4620874.
37. Moreira FA, Aguiar DC, Terzian AL, Guimarães FS, Wotjak CT. Cannabinoid type 1 receptors and transient receptor potential vanilloid type 1 channels in fear and anxiety-two sides of one coin? Neuroscience. 2012;204:186–92. https://doi.org/10.1016/j.neuroscience.2011.08.046.
38. Wang SE, Ko SY, Jo S, Choi M, Lee SH, Jo HR, Seo JY, Lee SH, Kim YS, Jung SJ, Son H. TRPV1 regulates stress responses through HDAC2. Cell Rep. 2017;19(2):401–12. https://doi.org/10.1016/j.celrep.2017.03.050.
39. Peiró AM, García-Gutiérrez MS, Planelles B, Femenía T, Mingote C, Jiménez-Treviño L, Martínez-Barrondo S, García-Portilla MP, Saiz PA, Bobes J, Manzanares J. Association of

cannabinoid receptor genes (*CNR1 and CNR2*) polymorphisms and panic disorder. Anxiety Stress Coping. 2020;33(3):256–65. https://doi.org/10.1080/10615806.2020.1732358.
40. Obermanns J, Meiser H, Hoberg S, Vesterager CS, Schulz F, Juckel G, Emons B. Genetic variation of the 5-HT1A rs6295, 5-HT2A rs6311, and CNR1 rs1049353 and an altered endocannabinoid system in depressed patients. Brain Behav. 2023;13(12):e3323. https://doi.org/10.1002/brb3.3323.
41. Marcus DJ, Bedse G, Gaulden AD, Ryan JD, Kondev V, Winters ND, Rosas-Vidal LE, Altemus M, Mackie K, Lee FS, Delpire E, Patel S. Endocannabinoid signaling collapse mediates stress-induced amygdalo-cortical strengthening. Neuron. 2020;105(6):1062–1076.e6. https://doi.org/10.1016/j.neuron.2019.12.024.
42. Longaretti A, Forastieri C, Gabaglio M, Rubino T, Battaglioli E, Rusconi F. Termination of acute stress response by the endocannabinoid system is regulated through lysine-specific demethylase 1-mediated transcriptional repression of 2-AG hydrolases ABHD6 and MAGL. J Neurochem. 2020;155(1):98–110. https://doi.org/10.1111/jnc.15000.
43. Dincheva I, Drysdale AT, Hartley CA, Johnson DC, Jing D, King EC, Ra S, Gray JM, Yang R, DeGruccio AM, Huang C, Cravatt BF, Glatt CE, Hill MN, Casey BJ, Lee FS. FAAH genetic variation enhances fronto-amygdala function in mouse and human. Nat Commun. 2015;6:6395. https://doi.org/10.1038/ncomms7395.
44. Geddes SD, Assadzada S, Lemelin D, Sokolovski A, Bergeron R, Haj-Dahmane S, Béïque JC. Target-specific modulation of the descending prefrontal cortex inputs to the dorsal raphe nucleus by cannabinoids. Proc Natl Acad Sci USA. 2016;113(19):5429–34. https://doi.org/10.1073/pnas.1522754113.
45. Lazary J, Lazary A, Gonda X, Benko A, Molnar E, Hunyady L, Juhasz G, Bagdy G. Promoter variants of the cannabinoid receptor 1 gene (CNR1) in interaction with 5-HTTLPR affect the anxious phenotype. Am J Med Genet B Neuropsychiatr Genet. 2009;150B(8):1118–27. https://doi.org/10.1002/ajmg.b.31024.
46. Cassano T, Gaetani S, Macheda T, Laconca L, Romano A, Morgese MG, Cimmino CS, Chiarotti F, Bambico FR, Gobbi G, Cuomo V, Piomelli D. Evaluation of the emotional phenotype and serotonergic neurotransmission of fatty acid amide hydrolase-deficient mice. Psychopharmacology. 2011;214(2):465–76. https://doi.org/10.1007/s00213-010-2051-0.
47. Lange MD, Daldrup T, Remmers F, Szkudlarek HJ, Lesting J, Guggenhuber S, Ruehle S, Jüngling K, Seidenbecher T, Lutz B, Pape HC. Cannabinoid CB1 receptors in distinct circuits of the extended amygdala determine fear responsiveness to unpredictable threat. Mol Psychiatry. 2017;22(10):1422–30. https://doi.org/10.1038/mp.2016.156.
48. García-Gutiérrez MS, García-Bueno B, Zoppi S, Leza JC, Manzanares J. Chronic blockade of cannabinoid CB2 receptors induces anxiolytic-like actions associated with alterations in GABA(A) receptors. Br J Pharmacol. 2012;165(4):951–64. https://doi.org/10.1111/j.1476-5381.2011.01625.x.
49. Ortega-Alvaro A, Aracil-Fernández A, García-Gutiérrez MS, Navarrete F, Manzanares J. Deletion of CB2 cannabinoid receptor induces schizophrenia-related behaviors in mice. Neuropsychopharmacology. 2011;36(7):1489–504. https://doi.org/10.1038/npp.2011.34.
50. Spanagel R. Cannabinoids and the endocannabinoid system in reward processing and addiction: from mechanisms to interventions. Dialogues Clin Neurosci. 2020;22(3):241–50. https://doi.org/10.31887/DCNS.2020.22.3/rspanagel.
51. Zhang HY, Gao M, Shen H, Bi GH, Yang HJ, Liu QR, Wu J, Gardner EL, Bonci A, Xi ZX. Expression of functional cannabinoid CB2 receptor in VTA dopamine neurons in rats. Addict Biol. 2017;22(3):752–65. https://doi.org/10.1111/adb.12367.
52. Ishiguro H, Iwasaki S, Teasenfitz L, Higuchi S, Horiuchi Y, Saito T, Arinami T, Onaivi ES. Involvement of cannabinoid CB2 receptor in alcohol preference in mice and alcoholism in humans. Pharmacogenomics J. 2007;7(6):380–5. https://doi.org/10.1038/sj.tpj.6500431.
53. Lupica CR, Riegel AC. Endocannabinoid release from midbrain dopamine neurons: a potential substrate for cannabinoid receptor antagonist treatment of addiction. Neuropharmacology. 2005;48(8):1105–16. https://doi.org/10.1016/j.neuropharm.2005.03.016.

54. Shen B, Zhang D, Zeng X, Guan L, Yang G, Liu L, Huang J, Li Y, Hong S, Li L. Cannabidiol inhibits methamphetamine-induced dopamine release via modulation of the DRD1-MeCP2-BDNF-TrkB signaling pathway. Psychopharmacology. 2022;239(5):1521–37. https://doi.org/10.1007/s00213-021-06051-y.
55. Pabalan N, Chaweeborisuit P, Tharabenjasin P, Tasanarong A, Jarjanazi H, Eiamsitrakoon T, Tapanadechopone P. Associations of CB1 cannabinoid receptor (CNR1) gene polymorphisms with risk for alcohol dependence: evidence from meta-analyses of genetic and genome-wide association studies. Medicine (Baltimore). 2021;100(43):e27343. https://doi.org/10.1097/MD.0000000000027343.
56. Chukwueke CC, Kowalczyk WJ, Gendy M, Taylor R, Tyndale RF, Le Foll B, Heishman SJ. The CB1R rs2023239 receptor gene variant significantly affects the reinforcing effects of nicotine, but not cue reactivity, in human smokers. Brain Behav. 2021;11(2):e01982. https://doi.org/10.1002/brb3.1982.
57. Gunasekera B, Diederen K, Bhattacharyya S. Cannabinoids, reward processing, and psychosis. Psychopharmacology. 2022;239(5):1157–77. https://doi.org/10.1007/s00213-021-05801-2.
58. Bioque M, Mas S, Costanzo MC, Cabrera B, Lobo A, González-Pinto A, Rodriguez-Toscano E, Corripio I, Vieta E, Baeza I, Ibáñez Á, Fraile MG, Cuesta MJ, Mezquida G, Lafuente A, Bernardo M, PEPs GROUP. Gene-environment interaction between an endocannabinoid system genetic polymorphism and cannabis use in first episode of psychosis. Eur Neuropsychopharmacol. 2019 Jun;29(6):786–94. https://doi.org/10.1016/j.euroneuro.2019.04.005.
59. Okahisa Y, Kodama M, Takaki M, Inada T, Uchimura N, Yamada M, Iwata N, Iyo M, Sora I, Ozaki N, Ujike H. Association study of two cannabinoid receptor genes, CNR1 and CNR2, with methamphetamine dependence. Curr Neuropharmacol. 2011;9(1):183–9. https://doi.org/10.2174/157015911795017191.
60. West ML, Sharif S. Cannabis and psychosis. Child Adolesc Psychiatr Clin N Am. 2023;32(1):69–83. https://doi.org/10.1016/j.chc.2022.07.004.
61. Marconi A, Di Forti M, Lewis CM, Murray RM, Vassos E. Meta-analysis of the association between the level of cannabis use and risk of psychosis. Schizophr Bull. 2016;42(5):1262–9. https://doi.org/10.1093/schbul/sbw003.
62. Giuffrida A, Leweke FM, Gerth CW, Schreiber D, Koethe D, Faulhaber J, Klosterkötter J, Piomelli D. Cerebrospinal anandamide levels are elevated in acute schizophrenia and are inversely correlated with psychotic symptoms. Neuropsychopharmacology. 2004;29(11):2108–14. https://doi.org/10.1038/sj.npp.1300558.
63. Bloomfield MA, Morgan CJ, Egerton A, Kapur S, Curran HV, Howes OD. Dopaminergic function in cannabis users and its relationship to cannabis-induced psychotic symptoms. Biol Psychiatry. 2014;75(6):470–8. https://doi.org/10.1016/j.biopsych.2013.05.027.
64. Leweke FM, Giuffrida A, Koethe D, Schreiber D, Nolden BM, Kranaster L, Neatby MA, Schneider M, Gerth CW, Hellmich M, Klosterkötter J, Piomelli D. Anandamide levels in cerebrospinal fluid of first-episode schizophrenic patients: impact of cannabis use. Schizophr Res. 2007;94(1–3):29–36. https://doi.org/10.1016/j.schres.2007.04.025.
65. Tao R, Li C, Jaffe AE, Shin JH, Deep-Soboslay A, Yamin R, Weinberger DR, Hyde TM, Kleinman JE. Cannabinoid receptor CNR1 expression and DNA methylation in human prefrontal cortex, hippocampus and caudate in brain development and schizophrenia. Transl Psychiatry. 2020;10(1):158. https://doi.org/10.1038/s41398-020-0832-8.
66. Hamdani N, Tabeze JP, Ramoz N, Ades J, Hamon M, Sarfati Y, Boni C, Gorwood P. The CNR1 gene as a pharmacogenetic factor for antipsychotics rather than a susceptibility gene for schizophrenia. Eur Neuropsychopharmacol. 2008;18(1):34–40. https://doi.org/10.1016/j.euroneuro.2007.05.005.
67. Hunt GE, Large MM, Cleary M, Lai HMX, Saunders JB. Prevalence of comorbid substance use in schizophrenia spectrum disorders in community and clinical settings, 1990-2017: systematic review and meta-analysis. Drug Alcohol Depend. 2018;191:234–58. https://doi.org/10.1016/j.drugalcdep.2018.07.011.

68. Ceccarini J, De Hert M, Van Winkel R, Peuskens J, Bormans G, Kranaster L, Enning F, Koethe D, Leweke FM, Van Laere K. Increased ventral striatal CB1 receptor binding is related to negative symptoms in drug-free patients with schizophrenia. NeuroImage. 2013;79:304–12. https://doi.org/10.1016/j.neuroimage.2013.04.052.
69. Banaszkiewicz I, Biala G, Kruk-Slomka M. Contribution of CB2 receptors in schizophrenia-related symptoms in various animal models: Short review. Neurosci Biobehav Rev. 2020;114:158–71. https://doi.org/10.1016/j.neubiorev.2020.04.020.

14 Reproduction, Embryogenesis, and Development

One of the last systems that medical cannabis providers typically think of as being influenced by the ECS is reproduction. However, it is actually an area in which the endocannabinoid system plays a very large role. Mammalian reproduction is controlled centrally by the hypothalamic-pituitary-gonadal (HPG) axis. This axis begins with the pulsatile excretion of gonadotropin-releasing hormone (GnRH) from the hypothalamus. In women, low-frequency pulses of GnRH excretion prompt release of follicle-stimulating hormone (FSH) from the pituitary, and high-frequency pulses cause luteinizing hormone (LH) to be released from the same area. The coordination of these two hormones then controls ovulation, maintenance of corpus luteum, and sex hormone synthesis and release. In men, the frequency of pulses is constant and drives spermatogenesis. As we will see, the endocannabinoid system plays a regulatory role at every step described [1].

14.1 ECS and Central Modulation of th e Hypothalamic-Pituitary-Gonadal Axis

GnRH-secreting neurons in the hypothalamus produce and secrete both 2-AG and AEA as well as expressing CB1 and CB2 receptors. AEA has been shown in cultured GnRH-secreting neurons to disrupt normal pulsatile secretion, most likely via CB1 activation [2]. This is supported by finding in mice that injection of AEA into the brain of male rats decreased serum LH and that injection of CB1 antagonist produced increased levels of LH, indicating tonic suppression by the ECS. On the contrary, female rats pretreated with estrogen actually had higher levels of LH produced [3]. Control of the ECS over GnRH cells can occur indirectly as well. GABAergic neurons have an excitatory effect on GnRH neurons. CB1 receptors on these GABAergic neurons inhibit GABA release and subsequently decrease GnRH release. Antagonism of these CB1 receptors allows greater GABA release, increasing GnRH release [4]. Chronic cannabis intake has long been known to decrease

J.-P. H. Dedam et al., *Diseases of the Endocannabinoid System*,
https://doi.org/10.1007/978-3-031-87484-0_14

male testosterone and spermatogenesis [5], as would be predicted by decreased LH levels. There is evidence as well that testosterone may have a feedback effect on the ECS as decreased testosterone levels in castrated rats cause decreased expression of CB1 receptor, although this was noted in the parotid gland not the HPG axis [6]. The regulation of the female reproductive system is significantly more complicated with estrogen performing a more nuanced feedback role. Rats still capable of producing estrogen had lower levels of CB1 expression in the hypothalamus than male rats or female mice that have had ovaries removed [7]. Estrogen was also able to decrease CB1 suppression of GABAergic neurons [8]. This would explain the above finding that female rats pretreated with estrogen were actually able to release more LH. While the exact mechanism of control of the HPG axis is not fully understood, it is clear that the ECS plays an important regulatory role. Levels of AEA and 2-AG fluctuate significantly in the brain throughout the menstrual cycle in rats, particularly in the hypothalamus and pituitary glands [9]. The ECS, therefore, appears to play an important role in the central regulation of the human reproductive system.

14.2 ECS in Testes and Ovaries

Moving peripherally, aside from the central modulation of hormones, the ECS plays a local role in the testes. Receptors are present in Leydig cells, interstitial testosterone-producing cells, and Sertoli cells that nourish developing sperm cells and germ cells themselves. The ECS appears necessary for the development of Leydig cells, and CB1 receptors rise and fall through the development process. Both CB1 and CB2 are necessary for the development of sperm cells. Both also affect motility in the mature sperm, with CB2 slowing sperm and CB1 immobilizing them [10]. It is, therefore, no surprise that one of the most consistent fertility suggestions for men is cannabis cessation.

As in the testes, the components of the ECS are well represented in the ovaries. The role of the ECS also appears to differ between species, so we will focus on humans. Serum AEA peaks increase during the follicular phase, peak in ovulation, then drop during the luteal phase, and correlate positively with estrogen, FSH, and LH but not progesterone [11]. CB1 and CB2 receptors are present on oocytes, and overactivation of CB2 can result in meiosis and impairs ovarian reserve in fetal gonads [12]. Oocytes develop into follicles that are composed of surrounding granulose cells, which produce different sex hormones throughout development of follicle; theca cells, which also produce hormones and fluids that nourish the oocyte; and the oocyte itself [13]. CB2 is expressed at higher levels than CB1 in the granulose cells and oocytes during the primordial, primary, and secondary follicular stages, with CB1 receptor expression increasing in the granulosa cells during the tertiary and corpus luteal phases. AEA also increases in concentration in the follicular fluid as the follicle develops [14].

14.3 ECS in Fallopian Tubes and Implantation and Pregnancy

After the release of the egg in ovulation, the egg transits down into the fallopian tubes where, in mice, appropriate AEA tone as regulated by expression of FAAH is important for fertilization and further transit of the egg [15]. Mouse studies also indicated that deficient CB1 expression in the fallopian tubes and endometrium affected transit. A study in women with ectopic pregnancy also showed lowered expression of CB1 in their fallopian tubes and endometrium than controls [16]. A more recent and larger study showed increased CB1 expression in the fallopian tubes of women with ectopic pregnancy, and a CB1 polymorphism, rs1049353, was found to be linked to higher CB1 expressions and ectopic pregnancy [17]. Proper implantation in the uterus also requires a precise gradient of both AEA and 2-AG in mice [15, 18]. Further discussion of the role of the ECS in blastocyte, embryo, and placental development will take place below. Spoiler alert, its important. AEA levels are higher in women during the first trimester than in the second and third trimesters with a significant spike during labor. These elevated levels are important for maintaining labor [19]. Is it possible that this spike in AEA may also help with pain and relaxation during labor and even account for the decreased memory of pain afterward. Lower serum levels of AEA have been found in women with preeclampsia [20]. Mice studies have also revealed that low expression of CB1 in mice leads to altered progesterone levels and preterm delivery [21]. Whether this occurs in women is something we should definitely be investigated.

14.4 ECS and PCOS

Polycystic ovarian syndrome (PCOS) is a common cause of infertility with known dysregulation of the ECS. Women with PCOS showed significant increases in serum AEA levels as well as lower endometrial FAAH expression. This alternation on FAAH expression improved with birth control and metformin [22]. As a disease of both metabolism and infertility due to hormonal imbalance, an ECS excess syndrome may play a larger role in pathophysiology than previously recognized as evidenced by polymorphisms in CB1 genes, RS 12720071 and RS 1049353, which were found to create a 3.6-fold increase in free androgen levels versus normal subjects. Elevated androgen levels are a known pathophysiologic feature of PCOS [23]. Fibroid development also appears to involve a localized endocannabinoid excessive syndrome as low FAAH levels lead to high AEA levels correlating to further growth [24]. Despite the obvious enormous influence of the ECS on human reproduction, unfortunately, this remains a very understudied area of medicine. As noted above, disruptions of the ECS can lead to infertility, ectopic pregnancy, PCOS, and fibroids. Hopefully, given the amount of money being spent on treatment of these conditions will prompt further investigation and understanding of the role of the ECS and how its modulation may lead to important treatments.

14.5 ECS in Embryo

Once fertilization of the egg occurs, it is termed as an embryo. Before implantation, it undergoes a number of mitotic divisions, first into two cells, then four, and then eight. It then compacts into a morula while continuing to move toward the uterus. A cavity develops inside, marking it now as a blastocyst. The cells continue dividing and form two distinct populations: the inner cell mass that will become the embryo and the outer layer that will become the placenta and extra-embryonic membranes. At this point, the blastocyst is capable of implantation. In mice, CB2 is expressed throughout the entire process, while CB1 is expressed starting at the two-cell stage and goes on to be located mainly in the outer layer of the blastocyst. NAPE-PLD, the enzyme that produces AEA, is present from the one-cell stage, and DAGL, which produces 2-AG, is detectable from day 1 but is not substantially unregulated until day 5 [25]. CB1 knockout mice embryos failed to develop and implant normally. Exogenous native cannabinoids administered during this stage also impeded development and prevented implantation. Tightly regulated levels of CB1 activation were therefore needed both for preimplantation development and implantation. CB2 knockout mice embryos also had impaired development [26].

14.6 ECS in Neurodevelopment

Through the rest of fetal development, the best studied area is neurodevelopment. Neural stem cells (NSC) begin proliferating as the number of neuron cells is critical to the formation of a normal neural net. Initially, they begin symmetrically dividing, producing more stem cells. Eventually, asymmetrical division occurs when one of the daughter cells, after several more rounds of proliferation, will begin to differentiate into different types of neurons. The ECS appears to play a role in this process as well. Studies of cannabinoid receptor density in developing human brains show relatively low levels up until 14 weeks before which much of the neural proliferation has already occurred and migration begins. Levels increase until reaching adult levels around 24 weeks of gestation; during this period of time, migration continues and synaptogenesis begins. Cannabinoid levels are not homogeneous, however with the lowest densities in the caudate and putamen and highest in the globus pallidus [27]. They are also found in higher amounts in white matter areas of the rat brain during development, but largely absent there in adults, indicating a role in the development of areas such as the corpus callosum and anterior commissure [28]. Activation of the CB1 receptor in mouse neural stem/proliferator cells for 48 h increased proliferation and after 1 week resulted in differentiation [29]. In the cerebellum, granule cell precursor cells, which we saw in Chap. 12, need CB1 receptor activation for proliferation via the AKT/glycogen synthase kinase-3beta/beta-catenin signaling [30]. Activation of CB2 receptor has also been shown to increase neural progenitor cell proliferation in embryotic cells via the mTORC1 pathways, but also in adult hippocampus cells [31]. The ECS plays a role in late-stage neural development as well. Studies in rats showed that CB1 receptor activation contributed to complex signaling pathways to generate deep layer cortical neurons and influenced connectivity to corticospinal motor neurons, the impairment of which resulted in abnormal skilled motor function in adult rats [32]. Aside from neurons, 2-AG drove oligodendrocyte differentiation via CB1- And CB-dependent pathways [33] as well as regulating their migration [34].

14.7 ECS in Postnatal Neurodevelopment and Autism and Fragile X Syndrome

Neural development continues postnatally as well in human males lasting up until 26 years of age. The ECS continues to influence neural development during this period of time as well. CB1 knockout mice have been shown to have impaired communication and loss of social interest/investigation during development and as adults. Previous studies on the same mice also showed repetitive and inflexible behavior. These findings lead the authors of the study to speculate that a disruption in the ECS during neural development could contribute to autism spectrum disorders [35]. Another syndrome of neurodevelopment known to be associated with ECS dysfunction is fragile X syndrome. In this syndrome, the absence of fragile X messenger ribonucleoprotein 1 (FMR1) protein disrupts normal ECS function in neurodevelopment, leading to overstimulation, desensitization, and internalization of presynaptic CB1 receptors. Fragile X syndrome is characterized by intellectual disability, hyperactivity, seizures, and problems with social interactions similar to autism [36]. Blocking maternal CB1 receptors with rimonabant during the immediate postpartum periods lead to decreased maternal behaviors of attachment and resulted in lower birthweight and decreased vocalizations. However, as adults, they appear to have overcompensated and became more socially active than controls with increased attention to rearing [37]. CB2 knockout mice did not show significant developmental changes but did manifest expected changes in immune response and bone metabolism later on in life [38]. Development of the striatum, specifically of striosome-dendron bouquets, which control dopaminergic cell activity through inhibitory synapses, is also formed during the immediate postnatal period. CB1 knockout mice were found to have markedly disorganized bouquets at postnatal day 11 and this persisted into adulthood [39]. A polymorphism in the gene for FAAH C385A, which results in higher than normal AEA levels, was studied in the context of human brain connectivity between different cortical lobes and limbic, subcortical, brain stem, and cerebellar structures in adolescent with still developing brains. They demonstrated significant changes with increases in connectivity in some areas and decreases in others, while others saw no change. Interestingly, previously noted increased connections between the prefrontal cortex and amygdala that decreased anxiety symptoms in adults had not yet formed in adolescents [40]. While the authors did not comment on clinical significance of this finding, it is obvious that the ECS plays a very important role in brain development throughout the prenatal, neonatal, and even up to adolescent development. It is unlikely given the shear complexity of brain development, with trillions of connections being made and broken, that we will ever fully understand the role of the ECS on its development, but it is clear that its effects are immense and likely influence many common neuro-pyschologic diseases.

14.8 ECS in Breast Milk

In mice and almost certainly humans, endocannabinoids are abundant in mother's milk and in the postnatal brain. CB1 receptor activation has long been known to influence neonatal feeding. Administration of the CB1 blocker rimonabant to developing mice results in failure to thrive [41]. While maternal milk endocannabinoid content has not been extensively studied, maternal nutrition and subsequent endocannabinoid metabolic precursor content has been studied. Dietary long-chain fatty acids such as arachidonic acid and DHA are essential for developing neonatal brains, specifically for production of AEA. In an experiment in mice, maternal undernutrition resulted in lower AEA in the hypothalamus of the neonate and subsequent lower weight gain [42]. Excessive AEA administered during lactation in mice results in overweight mice with significant fat accumulation and insulin resistant in adults along with higher levels of leptin [43]. Neonatal stress in mice also increased serum leptin and corticosteroid levels in the blood with increased abundance in CB1 receptor on the liver and adult insulin resistance [44]. As the ECS is unregulated during the stress response as seen in Chap. 12, it may also be playing a role in the adverse metabolic effects mediated by stress hormone as overactivation of the ECS was shown to produce in Chap. 4. Very-high-fat diets were also capable of overactivation the ECS in neonates, and these changes could be partially reversed by returning to a normal diet [45]. Despite these findings, no compelling studies in humans have found that cannabis exposure in the prenatal/postnatal period has significant effects on development. The extensive NASEM report on health effects of cannabis from 2017 showed only a tendency toward slightly lower birth weights in babies born to women with history of smoked cannabis use, which would not have been predicted by the above animal experiments (NASAM [46]).

References

1. Pařízek A, Suchopár J, Laštůvka Z, Alblová M, Hill M, Dušková M. The endocannabinoid system and its relationship to human reproduction. Physiol Res. 2023;72(S4):S365–80. https://doi.org/10.33549/physiolres.935229.
2. Gammon CM, Freeman GM Jr, Xie W, Petersen SL, Wetsel WC. Regulation of gonadotropin-releasing hormone secretion by cannabinoids. Endocrinology. 2005;146(10):4491–9. https://doi.org/10.1210/en.2004-1672.
3. Scorticati C, Fernández-Solari J, De Laurentiis A, Mohn C, Prestifilippo JP, Lasaga M, Seilicovich A, Billi S, Franchi A, McCann SM, Rettori V. The inhibitory effect of anandamide on luteinizing hormone-releasing hormone secretion is reversed by estrogen. Proc Natl Acad Sci USA. 2004;101(32):11891–6. https://doi.org/10.1073/pnas.0404366101.
4. Farkas I, Kalló I, Deli L, Vida B, Hrabovszky E, Fekete C, Moenter SM, Watanabe M, Liposits Z. Retrograde endocannabinoid signaling reduces GABAergic synaptic transmission to

gonadotropin-releasing hormone neurons. Endocrinology. 2010;151(12):5818–29. https://doi.org/10.1210/en.2010-0638.
5. Kolodny RC, Masters WH, Kolodner RM, Toro G. Depression of plasma testosterone levels after chronic intensive marihuana use. N Engl J Med. 1974;290(16):872–4. https://doi.org/10.1056/NEJM197404182901602.
6. Busch L, Sterin-Borda L, Borda E. Effects of castration on cannabinoid cb receptor expression and on the biological actions of cannabinoid in the parotid gland. Clin Exp Pharmacol Physiol. 2006;33(3):258–63. https://doi.org/10.1111/j.1440-1681.2006.04355.x.
7. Riebe CJ, Hill MN, Lee TT, Hillard CJ, Gorzalka BB. Estrogenic regulation of limbic cannabinoid receptor binding. Psychoneuroendocrinology. 2010;35(8):1265–9. https://doi.org/10.1016/j.psyneuen.2010.02.008.
8. Nguyen QH, Wagner EJ. Estrogen differentially modulates the cannabinoid- induced presynaptic inhibition of amino acid neurotransmission in proopiomelanocortin neurons of the arcuate nucleus. Neuroendocrinology. 2006;84(2):123–37. https://doi.org/10.1159/000096996. Epub 2006 Nov 9. PMID: 17106183.
9. Bradshaw HB, Rimmerman N, Krey JF, Walker JM. Sex and hormonal cycle differences in rat brain levels of pain-related cannabimimetic lipid mediators. Am J Physiol Regul Integr Comp Physiol. 2006;291(2):R349–58. https://doi.org/10.1152/ajpregu.00933.2005.
10. Sayed TS, Balasinor NH, Nishi K. Diverse role of endocannabinoid system in mammalian male reproduction. Life Sci. 2021;286:120035. https://doi.org/10.1016/j.lfs.2021.120035.
11. El-Talatini MR, Taylor AH, Konje JC. The relationship between plasma levels of the endocannabinoid, anandamide, sex steroids, and gonadotrophins during the menstrual cycle. Fertil Steril. 2010;93(6):1989–96. https://doi.org/10.1016/j.fertnstert.2008.12.033.
12. De Domenico E, Todaro F, Rossi G, Dolci S, Geremia R, Rossi P, Grimaldi P. Overactive type 2 cannabinoid receptor induces meiosis in fetal gonads and impairs ovarian reserve. Cell Death Dis. 2017;8(10):e3085. https://doi.org/10.1038/cddis.2017.496.
13. Cecconi S, Rapino C, Di Nisio V, Rossi G, Maccarrone M. The (endo)cannabinoid signaling in female reproduction: what are the latest advances? Prog Lipid Res. 2020;77:101019. https://doi.org/10.1016/j.plipres.2019.101019.
14. El-Talatini MR, Taylor AH, Elson JC, Brown L, Davidson AC, Konje JC. Localisation and function of the endocannabinoid system in the human ovary. PLoS One. 2009;4(2):e4579. https://doi.org/10.1371/journal.pone.0004579.
15. Wang H, Xie H, Guo Y, Zhang H, Takahashi T, Kingsley PJ, Marnett LJ, Das SK, Cravatt BF, Dey SK. Fatty acid amide hydrolase deficiency limits early pregnancy events. J Clin Invest. 2006;116(8):2122–31. https://doi.org/10.1172/JCI28621.
16. Horne AW, Phillips JA 3rd, Kane N, Lourenco PC, McDonald SE, Williams AR, Simon C, Dey SK, Critchley HO. CB1 expression is attenuated in fallopian tube and decidua of women with ectopic pregnancy. PLoS One. 2008;3(12):e3969. https://doi.org/10.1371/journal.pone.0003969.
17. Moudi B, Heidari Z, Asemi-Rad A, Mahmoudzadeh-Sagheb H, Sheibak N, Ghasemi M, Eslami S. Cannabinoid receptor type-1 and its correlation with CB1 gene polymorphism-1359G/A in ectopic pregnancy compared to the control group. J Obstet Gynaecol Res. 2021;47(4):1256–64. https://doi.org/10.1111/jog.14688.
18. Wang H, Xie H, Sun X, Kingsley PJ, Marnett LJ, Cravatt BF, Dey SK. Differential regulation of endocannabinoid synthesis and degradation in the uterus during embryo implantation. Prostaglandins Other Lipid Mediat. 2007;83(1–2):62–74. https://doi.org/10.1016/j.prostaglandins.2006.09.009.
19. Habayeb OM, Taylor AH, Evans MD, Cooke MS, Taylor DJ, Bell SC, Konje JC. Plasma levels of the endocannabinoid anandamide in women--a potential role in pregnancy maintenance and labor? J Clin Endocrinol Metab. 2004;89(11):5482–7. https://doi.org/10.1210/jc.2004-0681.
20. Molvarec A, Fügedi G, Szabó E, Stenczer B, Walentin S, Rigó J Jr. Decreased circulating anandamide levels in preeclampsia. Hypertens Res. 2015;38(6):413–8. https://doi.org/10.1038/hr.2015.20.

21. Wang H, Xie H, Dey SK. Loss of cannabinoid receptor CB1 induces preterm birth. PLoS One. 2008;3(10):e3320. https://doi.org/10.1371/journal.pone.0003320.
22. Cui N, Feng X, Zhao Z, Zhang J, Xu Y, Wang L, Hao G. Restored plasma anandamide and endometrial expression of fatty acid amide hydrolase in women with polycystic ovary syndrome by the combination use of Diane-35 and metformin. Clin Ther. 2017;39(4):751–8. https://doi.org/10.1016/j.clinthera.2017.02.007.
23. Jędrzejuk D, Laczmański L, Kuliczkowska J, Lenarcik A, Trzmiel-Bira A, Hirnle L, Dorobisz U, Milewicz A, Lwow F, Urbanovych A, Słoka N. Selected CNR1 polymorphisms and hyperandrogenemia as well as fat mass and fat distribution in women with polycystic ovary syndrome. Gynecol Endocrinol. 2015;31(1):36–9.
24. Ayakannu T, Taylor AH, Marczylo TH, Konje JC. New insights of uterine leiomyoma pathogenesis: endocannabinoid system. Med Sci Monit Basic Res. 2019;25:76–87. https://doi.org/10.12659/MSMBR.914019.
25. Sun X, Dey SK. Aspects of endocannabinoid signaling in periimplantation biology. Mol Cell Endocrinol. 2008;286(1–2 Suppl 1):S3–S11. https://doi.org/10.1016/j.mce.2008.01.002.
26. Paria BC, Song H, Wang X, Schmid PC, Krebsbach RJ, Schmid HH, Bonner TI, Zimmer A, Dey SK. Dysregulated cannabinoid signaling disrupts uterine receptivity for embryo implantation. J Biol Chem. 2001;276(23):20523–8. https://doi.org/10.1074/jbc.M100679200.
27. Biegon A, Kerman IA. Autoradiographic study of pre- and postnatal distribution of cannabinoid receptors in human brain. NeuroImage. 2001;14(6):1463–8. https://doi.org/10.1006/nimg.2001.0939.
28. Berrendero F, Sepe N, Ramos JA, Di Marzo V, Fernández-Ruiz JJ. Analysis of cannabinoid receptor binding and mRNA expression and endogenous cannabinoid contents in the developing rat brain during late gestation and early postnatal period. Synapse. 1999;33(3):181–91.
29. Xapelli S, Agasse F, Sardà-Arroyo L, Bernardino L, Santos T, Ribeiro FF, Valero J, Bragança J, Schitine C, de Melo Reis RA, Sebastião AM, Malva JO. Activation of type 1 cannabinoid receptor (CB1R) promotes neurogenesis in murine subventricular zone cell cultures. PLoS One. 2013;8(5):e63529. https://doi.org/10.1371/journal.pone.0063529.
30. Trazzi S, Steger M, Mitrugno VM, Bartesaghi R, Ciani E. CB1 cannabinoid receptors increase neuronal precursor proliferation through AKT/glycogen synthase kinase-3beta/beta-catenin signaling. J Biol Chem. 2010;285(13):10098–109. https://doi.org/10.1074/jbc.M109.043711.
31. Palazuelos J, Ortega Z, Díaz-Alonso J, Guzmán M, Galve-Roperh I. CB2 cannabinoid receptors promote neural progenitor cell proliferation via mTORC1 signaling. J Biol Chem. 2012;287(2):1198–209. https://doi.org/10.1074/jbc.M111.291294.
32. Díaz-Alonso J, Aguado T, Wu CS, Palazuelos J, Hofmann C, Garcez P, Guillemot F, Lu HC, Lutz B, Guzmán M, Galve-Roperh I. The CB(1) cannabinoid receptor drives corticospinal motor neuron differentiation through the Ctip2/Satb2 transcriptional regulation axis. J Neurosci. 2012;32(47):16651–65. https://doi.org/10.1523/JNEUROSCI.0681-12.2012.
33. Gomez O, Arevalo-Martin A, Garcia-Ovejero D, Ortega-Gutierrez S, Cisneros JA, Almazan G, Sánchez-Rodriguez MA, Molina-Holgado F, Molina-Holgado E. The constitutive production of the endocannabinoid 2-arachidonoylglycerol participates in oligodendrocyte differentiation. Glia. 2010;58(16):1913–27. https://doi.org/10.1002/glia.21061.
34. Sanchez-Rodriguez MA, Gomez O, Esteban PF, Garcia-Ovejero D, Molina-Holgado E. The endocannabinoid 2-arachidonoylglycerol regulates oligodendrocyte progenitor cell migration. Biochem Pharmacol. 2018;157:180–8. https://doi.org/10.1016/j.bcp.2018.09.006.
35. Fyke W, Premoli M, Echeverry Alzate V, López-Moreno JA, Lemaire-Mayo V, Crusio WE, Marsicano G, Wöhr M, Pietropaolo S. Communication and social interaction in the cannabinoid-type 1 receptor null mouse: implications for autism spectrum disorder. Autism Res. 2021;14(9):1854–72. https://doi.org/10.1002/aur.2562.
36. Palumbo JM, Thomas BF, Budimirovic D, Siegel S, Tassone F, Hagerman R, Faulk C, O'Quinn S, Sebree T. Role of the endocannabinoid system in fragile X syndrome: potential mechanisms for benefit from cannabidiol treatment. J Neurodev Disord. 2023;15(1):1. https://doi.org/10.1186/s11689-023-09475-z.

37. Schechter M, Pinhasov A, Weller A, Fride E. Blocking the postpartum mouse dam's CB1 receptors impairs maternal behavior as well as offspring development and their adult social-emotional behavior. Behav Brain Res. 2012;226(2):481–92. https://doi.org/10.1016/j.bbr.2011.10.016.
38. Buckley NE. The peripheral cannabinoid receptor knockout mice: an update. Br J Pharmacol. 2008;153(2):309–18. https://doi.org/10.1038/sj.bjp.0707527.
39. Crittenden JR, Yoshida T, Venu S, Mahar A, Graybiel AM. Cannabinoid receptor 1 is required for neurodevelopment of Striosome-Dendron bouquets. eNeuro. 2022;9(2) https://doi.org/10.1523/ENEURO.0318-21.2022.
40. Sisk LM, Rapuano KM, Conley MI, Greene AS, Horien C, Rosenberg MD, Scheinost D, Constable RT, Glatt CE, Casey BJ, Gee DG. Genetic variation in endocannabinoid signaling is associated with differential network-level functional connectivity in youth. J Neurosci Res. 2022;Mar;100(3):731–743. https://doi.org/10.1002/jnr.24946. Epub 2021 Sep 8. PMID: 34496065; PMCID: PMC8866205.
41. Fride E, Braun H, Matan H, Steinberg S, Reggio PH, Seltzman HH. Inhibition of milk ingestion and growth after administration of a neutral cannabinoid CB1 receptor antagonist on the first postnatal day in the mouse. Pediatr Res. 2007;62(5):533–6. https://doi.org/10.1203/PDR.0b013e3181559d42.
42. Matias I, Léonhardt M, Lesage J, De Petrocellis L, Dupouy JP, Vieau D, Di Marzo V. Effect of maternal under-nutrition on pup body weight and hypothalamic endocannabinoid levels. Cell Mol Life Sci. 2003;60(2):382–9. https://doi.org/10.1007/s000180300031.
43. Aguirre CA, Castillo VA, Llanos MN. Excess of the endocannabinoid anandamide during lactation induces overweight, fat accumulation and insulin resistance in adult mice. Diabetol Metab Syndr. 2012;4(1):35. https://doi.org/10.1186/1758-5996-4-35.
44. Fonseca L, Castillo V, Aguirre C, Silva P, Ronco AM, Llanos M. Stress during lactation induces insulin resistance associated with an increase in type 1 cannabinoid receptors in liver and adipose tissue. J Nutr Metab. 2019;2019:2806519. https://doi.org/10.1155/2019/2806519.
45. Barrera C, Castillo V, Valenzuela R, Valenzuela CA, Garcia-Diaz DF, Llanos M. Effects on fetal metabolic programming and endocannabinoid system of a Normocaloric diet during pregnancy and lactation of female mice with Pregestational obesity. Nutrients. 2023;15(16):3531. https://doi.org/10.3390/nu15163531.
46. National Academies of Sciences, Engineering, and Medicine; Health and Medicine Division; Board on Population Health and Public Health Practice; Committee on the Health Effects of Marijuana: An Evidence Review and Research Agenda. The Health Effects of Cannabis and Cannabinoids: The Current State of Evidence and Recommendations for Research. Washington (DC): National Academies Press 2017.

The Eye

15

This brief chapter will focus on the eye. One of the earliest well-known indications for medical cannabis was for treatment of glaucoma, although this represents only one part of the role of the ECS in the eye. The ECS is present in the majority of eye tissues including anterior structures that modulate interocular pressure but also the retina. CB1 is found in the ciliary body, the trabecular meshwork, Schlemm's canal, and the retina. CB2 may be present in anterior structures, and CB2 knockout mice studies indicate a role in normal vision, making its presence in the retina extremely likely as well. TRPV1 is also present in the retina, as is GPR18.

15.1 ECS and Intraocular Pressure

IOP is regulated by the rate of production versus the outflow of the aqueous humor, the clear liquid that gives the eye its shape and fills in the interior of the eye. It is produced in the ciliary bodies, and CB1 and GP18 receptors on the epithelial membrane may regulate active passage of solutes in the posterior chamber. Aqueous fluid is removed from the eye via two pathways. The trabecular network, located at the angle of cornea and iris, allows flow of fluid into Schlemm's canal and then exits through the episcleral vessel. Changes in ciliary body muscles and smooth muscle-like contractile elements of the tissue itself effect flow. Both CB1 and CB2 receptors are present here but only activation of the CB1 receptor is known to affect IOP. This role is mediated via beta-adrenergic receptors with beta-adrenergic agonist reducing aqueous fluid secretion and beta-adrenergic antagonist increasing aqueous humor outflow. GP18 agonists also lower IOP [1]. Studies have been done using cannabinoids to treat glaucoma, typically administered orally, using THC as well as synthetic cannabinoids such as dronabinol and nabilone. Results demonstrated a 10–30% reduction in IOP 2–4 h after dosing with a dose-dependent relationship between the amount of medication and lowering of IOP; however, dose increases did not increase the duration of action with patients having to take medicine six

J.-P. H. Dedam et al., *Diseases of the Endocannabinoid System*,
https://doi.org/10.1007/978-3-031-87484-0_15

times a day for effective treatment. Topical administration with powerful synthetic CB1 agonist is a promising route of administration given the lack of systemic psychoactive side effects, and a solid lipid nanoparticle-delivered THC prodrug (to increase water solubility) has showed promise [2]. A study looking at CB1 knockout mice has shown that this mice actually have lower IOP than wild-type mice. The authors of the study speculated that there existed the previously mentioned mechanism for lowering IOP with activation of CB1 receptors, but also another mechanism where decreased CB1 activation also lowered IOP. This is supported by finding that CB1 antagonist also lowers IOP under some conditions and raised it in others depending on light cycle and sex of test subject [3]. These findings make the likelihood of glaucoma being an endocannabinoid deficiency syndrome exceedingly small.

15.2 ECS in the Retina

As mentioned above, the components of the ECS are extensively expressed in the retina as well and are known to function in the typical feedback inhibition fashion, specifically on GABAergic and glutamatergic neurons. However, the only functional effect thus far elicited was in CB2 knockout mice that exhibited changes in light processing [4]. Studies in mice models of retinitis pigmentosa, a degenerative disease of the retina, showed that application of a potent agonist of both CB1 and CB2 preserved rod and cone function and likely played a neuroprotective role [5]. Endocannabinoids are known to be considerably unregulated in both diabetic retinopathy and macular degeneration [6], and studies using cannabinoid receptor 1 and 2 agonists protected retinal cells from oxidative damage [7]. A more recent study, however, showed that CB2 activation on its own actually increased cell death and inflammation in human retinal pigment epithelium model of macular degeneration [8]. This would represent perhaps the only instance where CB2 activation results in increased inflammation. A large population study in the UK showed that frequent cannabis users were much less likely to develop macular degeneration, presumably due to its inflammatory effects, while cigarette smoking is a known major risk factor. However, in the cannabis users that did develop macular degeneration, they developed it much early in life. The authors theorized that it was related to the anti-angiogenic properties; however, it is also possible that these patients had overactivation of CB2 receptor as well as due to an unrecognized polymorphism [9].

References

1. Cairns EA, Baldridge WH, Kelly ME. The endocannabinoid system as a therapeutic target in glaucoma. Neural Plast. 2016;2016:9364091. https://doi.org/10.1155/2016/9364091.
2. Lindner T, Schmidl D, Peschorn L, Pai V, Popa-Cherecheanu A, Chua J, Schmetterer L, Garhöfer G. Therapeutic potential of cannabinoids in glaucoma. Pharmaceuticals (Basel). 2023;16(8):1149. https://doi.org/10.3390/ph16081149.

3. Miller S, Daily L, Ploss M, Greig I, Ross R, Rayana NP, Dai J, Sugali CK, Mao W, Straiker A. Evidence that cannabinoid CB1 receptors regulate intraocular pressure via two opposing mechanisms. Exp Eye Res. 2020;200:108241. https://doi.org/10.1016/j.exer.2020.108241.
4. Cécyre B, Zabouri N, Huppé-Gourgues F, Bouchard JF, Casanova C. Roles of cannabinoid receptors type 1 and 2 on the retinal function of adult mice. Invest Ophthalmol Vis Sci. 2013;54(13):8079–90. https://doi.org/10.1167/iovs.13-12514.
5. Lax P, Esquiva G, Altavilla C, Cuenca N. Neuroprotective effects of the cannabinoid agonist HU210 on retinal degeneration. Exp Eye Res. 2014;120:175–85. https://doi.org/10.1016/j.exer.2014.01.019.
6. Matias I, Wang JW, Moriello AS, Nieves A, Woodward DF, Di Marzo V. Changes in endocannabinoid and palmitoylethanolamide levels in eye tissues of patients with diabetic retinopathy and age-related macular degeneration. Prostaglandins Leukot Essent Fatty Acids. 2006;75(6):413–8. https://doi.org/10.1016/j.plefa.2006.08.002.
7. Wei Y, Wang X, Wang L. Presence and regulation of cannabinoid receptors in human retinal pigment epithelial cells. Mol Vis. 2009 Jun 14;15:1243–51. PMID: 19547718; PMCID: PMC2697670.
8. Hytti M, Andjelic S, Josifovska N, Piippo N, Korhonen E, Hawlina M, Kaarniranta K, Nevalainen TJ, Petrovski G, Parkkari T, Kauppinen A. CB2 receptor activation causes an ERK1/2-dependent inflammatory response in human RPE cells. Sci Rep. 2017;7(1):16169. https://doi.org/10.1038/s41598-017-16524-w.
9. Lehrer S, Rheinstein PH. Cannabis smoking and age-related macular degeneration in the UK biobank cohort. J Fr Ophtalmol. 2022;45(7):756–61. https://doi.org/10.1016/j.jfo.2022.01.004.

Cell Death and Cancer

16

The endocannabinoid system is heavily involved in suppression of tumor cell proliferation, tumor invasion, metastasis, angiogenesis, induction of tumor cell apoptosis, and immune response. Many patients have attempted to self-treat their cancer, usually with extremely high doses of cannabinoids, based on this information. In this final chapter, we will explore probably the most controversial use of medicinal cannabis: cancer. While many oncologists have been among the first to openly advocate for the use of medical cannabis in treating the side effects of cancer and cancer treatment, their position on the potential of cannabis is decidedly much more guarded.

16.1 ECS Expression in Cancer

The ECS was first implicated in inhibition of cellular proliferation in the 1970s with a study demonstrating that THC could inhibit the growth of lung adenocarcinoma. As has been noted over and over in this book, cannabinoid receptors are present in almost every cell of the body, and this includes many cancer cells as well. Components of the ECS have also been studied in a great variety of cancers, with mixed levels of expression, which may have clinical significance in treatment. Acute myeloid leukemia has increased CB2 expression. B-cell lymphomas have been shown to have increased 2-AG concentrations. Various breast cancers have shown increases in CB2 expression. Colorectal cancer has shown increases in 2-AG and AEA concentrations, and decreases in CB1 have been noted in higher-grade cancers. Endometrial cancer has increases in both endocannabinoids and both receptors. Both receptors are also increased in thyroid cancers. Increased CB1 is seen in gastric and esophageal cancers. CB1 receptor expression is increased in ovarian, pancreatic, and prostate cancers but is decreased in renal cell cancer. Glioblastoma consistently shows increases in CB2 receptor expression as do hepatocellular and some lung cancers. Increased CB1 and CB2 receptor activation are seen in mantle

J.-P. H. Dedam et al., *Diseases of the Endocannabinoid System*,
https://doi.org/10.1007/978-3-031-87484-0_16

cell lymphoma [1]. While these changes in the ECS have been associated with either improved or worse prognosis, it is obvious that the ECS is involved in states of dysregulated cell growth like cancer.

16.2 The ECS and Control of Normal Cell Growth

The ECS is known to participate in a great number of signaling pathways that control normal cell growth as well as tumor growth through a variety of different mechanisms such as induction of autophagy or apoptosis. As can be imagined, the effects are often complicated. In a study of cloned human astrocytomas, low levels of expression of both CB1 and CB2 receptors induced apoptosis via the ERK1/2 pathways. High levels of expression of both receptors did not cause cell death and activated instead to the AKT pathway. High levels of exposure to cannabinoids was capable of inducing apoptosis independently of both receptors but still via the ERK1/2 signaling pathway [2]. TRPV, GPR55, and PPAR gamma have also been implicated in the regulation of cell survival. TRPV1 is overexpressed in prostate cancer and underexpressed in urothelial bladder cancer, where its activation leads to cell apoptosis. Similar findings in gliomas were found to be activated by the p38 signaling pathway [3]. ECS modulation of ceramide, a bioactive lipid signal that promotes apoptosis, has also been demonstrated, again with mixed results. In astrocytes, where the ECS serves a neuroprotective role, it prevented cell death induced by ceramide via the phosphatidylinositol 3-kinase/protein kinase B pathway [4]. In prostate cancer, AEA increases ceramide synthesis, leading to cell death [5]. Treatment of prostate cancer cells with a CB1 and CB2 synthetic agonist, WIN 55, also induced cell death via a ERK1/2 pathway [6]. And also in prostate cancer, AEA induced cell death via ASK1 and p38MAPK/JNK pathways [7]. In mantle cell carcinoma, increased ceramide also leads to cell death via a CB1-dependent mechanism [8] and in pancreatic cancer via CB2 receptor-dependent mechanism and downstream upregulation of the p8 signaling pathways and inhibition of the mTORC1 pathway [9]. In hepatocellular carcinoma, CB2 receptor activation induced autophagy via tribbles homolog 3 (TRB3) upregulation, inhibition of Akt/mToRC1 axis and AMPK stimulation [10]. The upregulation of CB2 noted in the previous paragraph in hepatocellular is likely a positive thing. Activation of cannabinoid receptors inhibits adenyl cyclase activity and activates MAPK, which leads to pro-apoptotic effects in a variety of different cancers. Again in hepatocellular carcinoma, a synthetic CB1 and CB2 agonist, WIN 55, was used to induce apoptosis via the JNK/p38 MAPK signaling pathway. This also increased expression of PPAR gamma, another endocannabinoid target, and the effect was partially blocked by a PPAR gamma inhibitor [11]. CB2 activation in human leukemia induced apoptosis via p38 MAPK activation as well. In neurons, AEA-induced apoptosis required silencing of Bio, a heat-shock protein, in a CB1-dependent manner that involved p53, p42/44, and p38 MAPK [12]. Given its interaction with so many different signaling pathways, across so many different cancer types, there is no doubt that the ECS plays a very important role in the inhibition of cell growth.

16.3 ECS and Other Antineoplastic Mechanisms

The ECS is also involved in other potential antineoplastic processes such as cell cycle arrest, prevention of metastasis, and antiangiogenesis. Studies in breast cancer have shown that THC can lead to cell cycle arrest in G2-M phase by downregulation of cdc-2 [13]. In a separate study in breast cancer, AEA arrested cells in the S phase via Chk1 activation, Cdc25A degradation, and suppression of Cdk2 [14]. In a study in non-small lung cancer cells, THC was able to prevent EGF-induced growth and metastasis, likely via prevention of EGF-driven phosphorylation of ERK1/2, JNK1/2 and AKT [15]. CB1 and CB2 activation in mouse models of breast cancer arrested cancer cells in the G0–G1 phase but also inhibited angiogenesis via down-regulation of COX-2 pathway, which has been shown in other studies to confer resistance to apoptosis and increase angiogenesis and invasion when overexpressed in cancer cells [16]. In a model of thyroid cancer, AEA was found to inhibit bFGF endothelial cell proliferation, leading to apoptosis, but also inhibited matrix metalloprotease MMP-2 activity and angiogenesis [17]. In a separate study in gliomas, THC again inhibited matrix metalloproteases but also decreased expression of VEGF [18]. Decreased expression of VEGF along with placental growth factor and agiopoietin-2 was also seen in non-melanoma skin cancer cells treated with a synthetic CB1-CB2 agonist, WIN 55 [19]. Metastasis via the blood stream relies on expression of adhesion molecules. AEA has been shown to inhibit adhesion and migration along collagen of breast cancer cells. AEA-treated cells decreased phosphorylation of two tyrosine kinases, i.e., FAK and Src, via CB1 activation, which the authors of the study thought was the mechanism behind the decreased adhesion and migration [20].

16.4 Detrimental Effects of ECS in Cancer

This potentially beneficial finding is to be balanced with possibility that excessive endocannabinoid tone could be detrimental in cancer by limiting normal antitumor immune response. THC was shown in a mouse model of breast cancer to increase overall growth and metastasis, presumably via the IL-4- and IL-10-dependent, and CB2 receptor-mediated, shift from TH1 helper cells to TH2, thus decreasing the targeted antitumor activity. These are supported by findings of increased cancer remission and mortality in patients being treated with checkpoint inhibitors and also using moderate amounts of medical cannabis [21]. Checkpoint inhibitors function in cancer treatment but allow the immune system to ramp up to fight cancer. As was seen in Chap. 9 on the immune system, the endocannabinoid system in general blunts this response. The endocannabinoid 2-AG has also shown to have pro-invasive effects when administered exogenously to prostate cells, based on its rapid metabolism down to arachidonic acid, rather than its innate function [22].

The endocannabinoid system, as demonstrated above, is an important regulator of not only the cell cycle but also many other downstream factors that affect tumor growth such as angiogenesis, tissue invasion, and immune response. Modulation of the endocannabinoid system remains an active area of investigation within cancer research. There do not appear to be a studies tying polymorphism of the ECS to cancer. Given the degree to which the ECS is involved in the regulation of cell apoptosis, it would not be impossible that alteration in ECS function could result in either increased or decreased risk of cancer development. And as noted before, despite patients actively using high-dose cannabis to treat their cancers on their own, there are no randomized control trials as of yet evaluating the effectiveness of such an approach. This author is only aware of an ongoing study in the treatment of glioblastoma with cannabinoids used as an adjuvant therapy.

References

1. Garcia-Arencibia M, Molina-Holgado E, Molina-Holgado F. Effect of endocannabinoid signalling on cell fate: life, death, differentiation and proliferation of brain cells. Br J Pharmacol. 2019;176(10):1361–9. https://doi.org/10.1111/bph.14369.
2. Cudaback E, Marrs W, Moeller T, Stella N. The expression level of CB1 and CB2 receptors determines their efficacy at inducing apoptosis in astrocytomas. PLoS One. 2010;5(1):e8702. https://doi.org/10.1371/journal.pone.0008702.
3. Brown I, Cascio MG, Rotondo D, Pertwee RG, Heys SD, Wahle KW. Cannabinoids and omega-3/6 endocannabinoids as cell death and anticancer modulators. Prog Lipid Res. 2013;52(1):80–109. https://doi.org/10.1016/j.plipres.2012.10.001.
4. Gómez Del Pulgar T, De Ceballos ML, Guzmán M, Velasco G. Cannabinoids protect astrocytes from ceramide-induced apoptosis through the phosphatidylinositol 3-kinase/protein kinase B pathway. J Biol Chem. 2002;277(39):36527–33. https://doi.org/10.1074/jbc.M205797200.
5. Mimeault M, Pommery N, Wattez N, Bailly C, Hénichart JP. Anti-proliferative and apoptotic effects of anandamide in human prostatic cancer cell lines: implication of epidermal growth factor receptor down-regulation and ceramide production. Prostate. 2003;56(1):1–12. https://doi.org/10.1002/pros.10190.
6. Sarfaraz S, Afaq F, Adhami VM, Malik A, Mukhtar H. Cannabinoid receptor agonist-induced apoptosis of human prostate cancer cells LNCaP proceeds through sustained activation of ERK1/2 leading to G1 cell cycle arrest. J Biol Chem. 2006;281(51):39480–91. https://doi.org/10.1074/jbc.M603495200.
7. Sarker KP, Biswas KK, Yamakuchi M, Lee KY, Hahiguchi T, Kracht M, Kitajima I, Maruyama I. ASK1-p38 MAPK/JNK signaling cascade mediates anandamide-induced PC12 cell death. J Neurochem. 2003;85(1):50–61. https://doi.org/10.1046/j.1471-4159.2003.01663.x.
8. Gustafsson K, Sander B, Bielawski J, Hannun YA, Flygare J. Potentiation of cannabinoid-induced cytotoxicity in mantle cell lymphoma through modulation of ceramide metabolism. Mol Cancer Res. 2009;7(7):1086–98. https://doi.org/10.1158/1541-7786.MCR-08-0361.
9. Salazar M, Carracedo A, Salanueva IJ, Hernández-Tiedra S, Lorente M, Egia A, Vázquez P, Blázquez C, Torres S, García S, Nowak J, Fimia GM, Piacentini M, Cecconi F, Pandolfi PP, González-Feria L, Iovanna JL, Guzmán M, Boya P, Velasco G. Cannabinoid action induces autophagy-mediated cell death through stimulation of ER stress in human glioma cells. J Clin Invest. 2009;119(5):1359–72. https://doi.org/10.1172/jci37948.
10. Vara D, Salazar M, Olea-Herrero N, Guzmán M, Velasco G, Díaz-Laviada I. Anti-tumoral action of cannabinoids on hepatocellular carcinoma: role of AMPK-dependent activation of autophagy. Cell Death Differ. 2011;18(7):1099–111. https://doi.org/10.1038/cdd.2011.32.
11. Giuliano M, Pellerito O, Portanova P, Calvaruso G, Santulli A, De Blasio A, Vento R, Tesoriere G. Apoptosis induced in HepG2 cells by the synthetic cannabinoid WIN: involvement of the transcription factor PPAR gamma. Biochimie. 2009;91(4):457–65. https://doi.org/10.1016/j.biochi.2008.11.003.

12. Pasquariello N, Catanzaro G, Marzano V, Amadio D, Barcaroli D, Oddi S, Federici G, Urbani A, Finazzi Agrò A, Maccarrone M. Characterization of the endocannabinoid system in human neuronal cells and proteomic analysis of anandamide-induced apoptosis. J Biol Chem. 2009;284(43):29413–26. https://doi.org/10.1074/jbc.M109.044412.
13. Caffarel MM, Sarrió D, Palacios J, Guzmán M, Sánchez C. Delta9-tetrahydrocannabinol inhibits cell cycle progression in human breast cancer cells through Cdc2 regulation. Cancer Res. 2006;66(13):6615–21. https://doi.org/10.1158/0008-5472.CAN-05-4566.
14. Laezza C, Pisanti S, Crescenzi E, Bifulco M. Anandamide inhibits Cdk2 and activates Chk1 leading to cell cycle arrest in human breast cancer cells. FEBS Lett. 2006;580(26):6076–82. https://doi.org/10.1016/j.febslet.2006.09.074.
15. Preet A, Ganju RK, Groopman JE. Delta9-tetrahydrocannabinol inhibits epithelial growth factor-induced lung cancer cell migration in vitro as well as its growth and metastasis in vivo. Oncogene. 2008;27(3):339–46. https://doi.org/10.1038/sj.onc.1210641.
16. Qamri Z, Preet A, Nasser MW, Bass CE, Leone G, Barsky SH, Ganju RK. Synthetic cannabinoid receptor agonists inhibit tumor growth and metastasis of breast cancer. Mol Cancer Ther. 2009;8(11):3117–29. https://doi.org/10.1158/1535-7163.MCT-09-0448.
17. Pisanti S, Borselli C, Oliviero O, Laezza C, Gazzerro P, Bifulco M. Antiangiogenic activity of the endocannabinoid anandamide: correlation to its tumor-suppressor efficacy. J Cell Physiol. 2007;211(2):495–503. https://doi.org/10.1002/jcp.20954.
18. Blázquez C, Casanova ML, Planas A, Gómez Del Pulgar T, Villanueva C, Fernández-Aceñero MJ, Aragonés J, Huffman JW, Jorcano JL, Guzmán M. Inhibition of tumor angiogenesis by cannabinoids. FASEB J. 2003;17(3):529–31. https://doi.org/10.1096/fj.02-0795fje.
19. Casanova ML, Blázquez C, Martínez-Palacio J, Villanueva C, Fernández-Aceñero MJ, Huffman JW, Jorcano JL, Guzmán M. Inhibition of skin tumor growth and angiogenesis in vivo by activation of cannabinoid receptors. J Clin Invest. 2003;111(1):43–50. https://doi.org/10.1172/JCI16116.
20. Grimaldi C, Pisanti S, Laezza C, Malfitano AM, Santoro A, Vitale M, Caruso MG, Notarnicola M, Iacuzzo I, Portella G, Di Marzo V, Bifulco M. Anandamide inhibits adhesion and migration of breast cancer cells. Exp Cell Res. 2006;312(4):363–73. https://doi.org/10.1016/j.yexcr.2005.10.024.
21. Bar-Sela G, Cohen I, Campisi-Pinto S, Lewitus GM, Oz-Ari L, Jehassi A, Peer A, Turgeman I, Vernicova O, Berman P, Wollner M, Moskovitz M, Meiri D. Cannabis consumption used by cancer patients during immunotherapy correlates with poor clinical outcome. Cancers (Basel). 2020;12(9):2447. https://doi.org/10.3390/cancers12092447.
22. Endsley MP, Aggarwal N, Isbell MA, Wheelock CE, Hammock BD, Falck JR, Campbell WB, Nithipatikom K. Diverse roles of 2-arachidonoylglycerol in invasion of prostate carcinoma cells: Location, hydrolysis and 12-lipoxygenase metabolism. Int J Cancer. 2007;Sep;1;121(5):984–91. https://doi.org/10.1002/ijc.22761. PMID: 17443494; PMCID: PMC2565646.

Index

J.-P. H. Dedam et al., *Diseases of the Endocannabinoid System*,
https://doi.org/10.1007/978-3-031-87484-0

The manufacturer's authorised representative in the EU is Springer Nature Customer Service Centre GmbH, Europaplatz 3, 69115 Heidelberg, Germany. If you have any concerns regarding our products, please contact ProductSafety@springernature.com

Printed and bound by CPI Group (UK) Ltd, Croydon, CR0 4YY
01/10/2026
02225980-0006